D0215275

Physics of the Human Body

Physics of the Human Body

RICHARD P. MCCALL

The Johns Hopkins University Press

Baltimore

Printed in the United States of America on acid-free paper
9 8 7 6 5 4 3 2

The Johns Hopkins University Press
2715 North Charles Street
Baltimore, Maryland 21218-4363
www.press.jhu.edu

Library of Congress Cataloging-in-Publication Data

McCall, Richard Powell, 1955–
Physics of the human body / Richard P. McCall.
p. cm.
Includes bibliographical references and index.
ISBN-13: 978-0-8018-9455-8 (hardcover : alk. paper)
ISBN-10: 0-8018-9455-7 (hardcover : alk. paper)
ISBN-13: 978-0-8018-9456-5 (pbk. : alk. paper)
ISBN-10: 0-8018-9456-5 (pbk. : alk. paper)
1. Human physiology. I. Title.
QP34.5.M3564 2010
612—dc22 2009028860

A catalog record for this book is available from the British Library.

Contents

Preface vii

1 Motion and Balance 1

2 Fluids and Pressure 34

3 Energy, Work, and Metabolism 69

4 Sound, Speech, and Hearing 103

5 Electrical Properties and Cell Potential 136

6 Optics of the Eye 175

7 Biological Effects of Nuclear Radiation 210

8 Drug Delivery and Concentration 245

Notes 287

Index 291

Preface

What does physics have to do with the human body?" "I'm going into the medical field. Why do I need to take physics?" "Isn't physics all about math?" Perhaps you have asked these questions if you have taken a physics course or if you will have to study it in the future.

Often we associate physics with objects that move, such as baseballs and cars, or with the engineering disciplines and mathematics. It is true that physicists have a special fondness for the motion of objects. Pick up just about any high school or university physics book and you will see topics such as forces, linear and rotational motion, heat and thermodynamics, vibrations and waves, electricity and magnetism, optics, and quantum mechanics. These books go into great detail and include numerous applications, but these are mostly to inanimate objects.

There is, however, a growing desire among physicists to explore and discuss the relevance of physics to other disciplines. The physics of the human body is a topic of interest to many people. It is only in the past few years that a number of applications and examples related to the human body have begun to be included in textbooks.

The purpose of this book is to use basic physics principles to describe numerous actions and properties of the human body. Being on the faculty at a college of pharmacy, where every student is a pharmacy major, I have intentionally set out to teach a physics course that touches on some of the other disciplines, particularly chemistry, anatomy, and physiology. In addition, I have looked for ways to make connections to some of the upper-level science courses in the pharmacy curriculum, particularly in an area called pharmaceutics.

The typical college physics book and physics course go into quite a lot of mathematical detail. This book is not intended as a regular textbook for such a course, as it is much more descriptive in nature. For this book I go through some of the basic mathematics used and spend most of the time discussing how physics principles apply to the human body. The math can be instructive, however, and is intended to show some of the relationships between various parameters. For example, in the chapter on fluids, a discussion of Poiseuille's law helps us to understand that if a nearly completely blocked coronary artery can be opened by as little as 20% more, the blood flow should more than double.

The book is written at a level that a curious high school science student could understand. There is a minimum of mathematics, but enough that the student should be able to explore these ideas further. A university or college student interested in medical applications could use this book to understand concepts that may be considered in anatomy or physiology. And someone who is a practicing medical professional—a physician, a nurse, a pharmacist, a physical therapist, or an exercise physiologist—who wants to understand more about the underlying physics principles of the human body would likely find this book helpful.

The first chapter discusses motion of the human body and describes how the use of forces and torques allows a person to walk, to lift an object, or to maintain balance. We look at the importance of muscles, bones, and joints in the process of human motion. In chapter 2 we discuss fluids and pressure in the body, which leads naturally into a description of blood pressure, the heart that provides the pressure, and blood flow through arteries and veins. We look at pressure in various organs such as the lungs, the brain, the bladder, and the eye. In addition we discuss pressure applied to solids such as bones, with a brief look at bone density.

Chapter 3 describes energy in the body, including discussions on work, temperature and heat, and metabolism. Other topics include regulation of body temperature, food calories, and exercise. In chapter 4 we discuss principles of sound applied to speech and hearing. We look at waves and harmonic motion, the sound frequency spectrum, the production of speech in the larynx, and the detection of sound in the ear. We include a discussion of hearing problems and other applications of sound waves, such as ultrasound imaging.

In chapter 5 we turn our attention to electrical properties of the body.

We look at electric forces and fields, as well as electric energy and voltage. Static electricity, conduction of electrical signals along neurons, and a description of the membrane potential are included, as well as some applications such as the electrocardiogram or EKG. Chapter 6 focuses on optics of the eye; we discuss image formation by lenses in general, and specifically how the eye causes light to refract and form an image on the retina. We also describe vision problems and techniques for correcting them, such as contacts or LASIK.

Biological effects of nuclear radiation are the subject of chapter 7. In this chapter we describe the nucleus of the atom and look at nuclear radiation and its damaging effects on human tissue. But we also look at important useful applications of radiation, particularly in the medical field, such as diagnostic imaging and methods of treating cancer. In the last chapter, chapter 8, we look at a topic that is not part of a typical physics course, not even the one that I teach. The topic is drug concentrations in the body, particularly related to methods of taking medication and the elimination of medication from the body. This last chapter has quite a bit more mathematics in it, but its main focus is to discuss changes in concentration over time. It is an example of mathematical modeling that is often used by scientists to describe physical phenomena.

Most of the chapters in the book are independent and can be read in any order. There are occasional references to other chapters as attempts are made to link topics and ideas together, so there may be a need to refer to them to help gain more complete understanding.

I would like to thank a number of people who have helped with this book and have helped me to make the connections to other science disciplines. I wish to thank St. Louis College of Pharmacy (StLCoP) student Eric Venker for his careful reading of the manuscript and many helpful comments. (Eric is one of the top students I have ever had in a physics class.) Good discussions about anatomy, physiology, chemistry, and drugs were held with Dayton Ford, Marlene Katz, Leonard Naeger, Lucia Tranel, and Margaret Weck. Rasma Chereson and Theresa Laurent have helped me to understand drug concentrations in the body. Thanks go to current dean Kimberly Kilgore, who provided me with time and resources to write this book, and to former dean Ken Kirk, who was an especially strong supporter of my work here at StLCoP.

ONE

Motion and Balance

One of the most basic functions of the body is movement. We often see people in motion before we are able to see their eyes or make out the color of their hair. We may recognize people by the way they walk or how they hold their head. A wagging finger or a shaking head can be used to communicate. A football player may make a quick move to try to get additional yardage during a play. A basketball player can give a shake of the head to confuse an opponent about the direction of movement, but then quickly change course to shoot the ball. Unborn babies move inside their mothers. As I sit here in my chair, I am typing on the computer—my fingers are moving around all over the keyboard, my eyes are moving, I reach for the mouse, I pause to think, and I breathe and sneeze and yawn.

We exhibit so many different types of motion: we stand up, sit down, walk across the room, lift books, wave goodbye, throw baseballs, kick soccer balls, push and pull on objects, skate on ice, lean against walls, lie down, turn over, and on and on. What gives us the ability to move? How must our bodies act and react? And how do we keep from moving if we want to remain still? In this chapter we will look at motion of the human body, not motion of stuff that goes on inside the body like blood flow or nerve conduction (a microscopic view), but motion of the body or parts of the body as a whole (a macroscopic view). We will examine the forces and torques needed for movement and balance, and describe important physiological components of the body such as muscles, bones, and joints.

Forces

Let's start by looking at one of the most important factors in motion: force. A force is required to change the state of motion of an object. If an object is not moving and we want to get it to move, a force must be applied to the object. If an object is already moving and we want it to stop, then again we must apply a force to it.

Suppose you want to play catch with a friend using a baseball. The baseball is at rest on the ground, so you pick it up by applying a force to it. To throw the ball you must swing your arm around with the ball in your hand and then release it toward your friend. During the time the ball is in your hand, you are applying a force to it to change its state of motion. After you release the ball, it travels through the air, but not in a straight line; it rises, reaching some maximum height, and then falls, curving downward during the entire motion. All of this action begins as soon as you let go of the ball. Again, its state of motion is changing, so there is a force acting on it; that force is gravity, pulling it downward. To catch the ball, your friend must apply a force to stop it from moving, that is, to change its state of motion.

Not all forces change the motion of an object; some may actually keep an object from moving. For example, we don't typically want bridges to move. We want them to stay at rest, and especially not to collapse. So we build them in such a way that forces act at the right locations on the bridge to keep it in place.

Newton's Laws

There are several rules or laws that help us to describe these ideas. These are called Newton's laws of motion. You may be familiar with them already. I'll mention the first two here and bring up the third one in a bit.

Newton's first law of motion says that if an object is at rest or if it is moving, it will remain at rest or will continue moving with constant speed and direction as long as there is no net force acting on it. The first law is also called the law of inertia. *Inertia* is the tendency of an object to remain at rest or in its current state of motion. A measure of inertia is mass—if an object has a lot of mass then it has a lot of inertia; there is a strong tendency for it to remain at rest or to continue moving. But an object that has a small mass does not have a lot of inertia; it is fairly easy to get this object to move or to stop it from moving.

Consider a beach ball and a bowling ball. I can toss a beach ball to someone with little effort and it can be caught or knocked away very easily because its mass is small and it doesn't have much inertia. But it would be difficult for me to toss a bowling ball, and if I could toss it, it would be difficult for another person to catch it because its mass is large and it has a lot of inertia. In fact, the small mass of the beach ball is what makes it fun to play with—it is easy to throw, easy to catch, easy to hit, and so light that even the wind can blow it around. The large mass of the bowling ball makes it fun to use too (for bowling anyway); it may be hard to get it moving down the lane at the bowling alley, but when it crashes into the pins and sends them flying the result can be quite thrilling.

Newton's second law describes what happens when we apply a force to an object; not just any force, but something called *net force*. Net force is actually the sum (or difference) of all the forces acting on an object. If only one force acts on an object, then that one force is the net force. But if several forces are acting, then to get the net force, the individual forces must be added in a way that takes into account their directions of action. For example, suppose your car is stuck on the side of the road. If you and a friend both push in the same direction on the car, then your forces will add together. However, usually friction acts on the car in the opposite direction, so that force must be subtracted from the other two (fig. 1.1).

The second law states that, if a net force acts on an object, the object will accelerate, which means its state of motion will change. If the object is at rest, it will begin to move in the direction of the net force. For the state of motion of a moving object to change, it must speed up, slow down, or change direction. Newton's second law is also called the law of acceleration.

We can illustrate this by using the example of the car stuck on the side of the road. If you and your friend's forces combined are greater than the friction force, you will be able to get the car moving. Once it is moving, if you stop pushing on the car, but friction is still acting on it, it will slow down and stop.

Another consequence of Newton's second law is that if all the forces

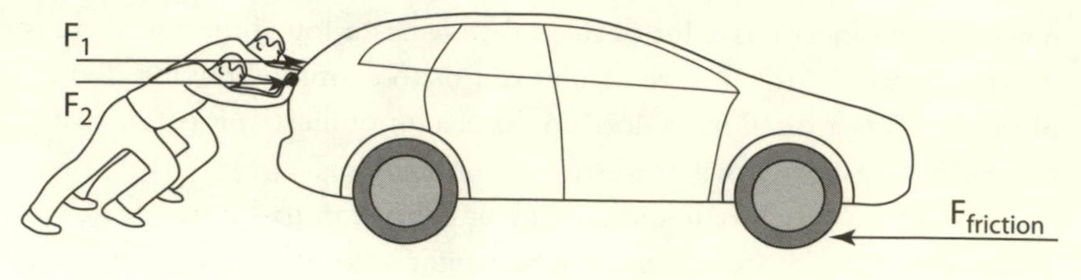

FIGURE 1.1. Net force. The net force F_{net} on the car is found by adding the two push forces F_1 and F_2 and subtracting the friction force $F_{friction}$.

acting on an object cancel each other out, then the net force is zero and the object will not accelerate—its state of motion will not change. If you and your friend get the car moving by pushing on it with a force greater than friction, you can get it to start moving. If you want to keep it moving, you don't have to push quite as hard because your two forces only need to match the frictional force exactly.

Let's consider the baseball again. When the ball is on the ground, the weight of the ball is matched by an upward force of the ground on the ball giving a net force of zero and the ball remains at rest. When you pick up the ball, the force you apply is greater than the weight of the ball, so you are able to get it to move upward. In order to throw the ball, you have to apply enough force to it in the proper direction for it to speed up quickly so it can fly through the air. At the same time, you have to overcome the gravitational force. Once the ball is airborne, gravity acts to pull it downward and air resistance acts to slow it down slightly. For your friend to catch the ball, he or she must apply a large force (overcoming any gravitational force as well) to slow it down quickly and bring it to a stop.

We can make Newton's second law more quantitative. The acceleration of the object depends on the net force acting on the object, but it also depends on the mass of the object. For more massive objects, the acceleration is smaller, and for objects that don't have much mass the acceleration is larger. If the car that you and your friend are pushing is a small compact, the acceleration will be greater than for a large full-size car.

Newton's second law can be written mathematically as

$$a = \frac{F_{net}}{m}, \qquad (1)$$

where a is the acceleration, F_{net} is the net force, and m is the mass of the object. This expression says that the acceleration of an object is proportional to the net force acting on the object and inversely proportional to the mass of the object. When I punch a beach ball it may have a large acceleration because its mass is so small. If that same force were applied to a bowling ball, its acceleration would be much smaller because its mass is quite large. To get a bowling ball to accelerate quickly, I must apply a much larger force to it because of its large mass. These ideas will come up from time to time as we discuss motion of the body.

Action-Reaction Pairs of Forces

Before getting to Newton's third law, I'll introduce an interesting feature of forces. If a force acts on an object, then the force must be caused by another object. In other words, for a force to exist there must be two objects involved—one object exerts the force and the other object has the force exerted on it. Sometimes it is easy to determine the two objects involved but at other times it is a bit more challenging. If you pick up a baseball, then you exert the force on the baseball. If gravity acts on the baseball, then the gravitational force is produced by the earth. If air resistance acts on the baseball, air exerts the force on the baseball. If your friend catches the ball, then the friend (or the baseball glove) exerts a force on the baseball. All of these forces are acting on the baseball and are produced by other objects—you, the earth, the air, and your friend.

Think again about pushing the car. There are several forces acting on the car—the push force, the opposing frictional force, the gravitational force (or weight), and the force holding the car up. You exert the push force, the ground and gears and bearings exert the frictional force, the earth exerts the gravitational force, and the ground exerts the force that holds the car up. When using Newton's first and second laws, we are mainly interested in the forces acting on the car, but sometimes we have to consider the object or objects that exert the forces.

Newton's third law considers not just the forces acting on an object, but also the effect on the object that exerts the force. It turns out that if one object exerts a force on another, then the second object exerts a force on the first. These two forces are equal and opposite. One force is considered the action force and the other is the reaction to it, so they are called an *action-reaction pair*. If you push on a car, then the car pushes back on you. If you kick a football by applying a force to the football with your foot, then the football pushes back on your foot. If you kick it many times, your foot will be tired and perhaps sore.

But if all forces occur in pairs that are equal and opposite in direction, don't they just cancel out? How can an object move? The key here is that when we consider the motion of an object, we focus on the forces acting on that object. Yes, that object will exert forces back on the other objects, but if we need to know what is happening to them, then we will focus on the forces acting on them. Each object will have forces acting on it and to study its motion, we need to look only at the forces acting on the object.

I bring up the idea of action-reaction pairs of forces because it is important in motion of the body. Whenever a part of the body moves, for example if you lift your forearm, a muscle pulls on that body part, the biceps muscle in that case. The reaction to this force is the body part pulling on the muscle. But there is another force acting on the muscle to keep it from moving, because muscles are attached at both ends. So if one end is attached to a body part and is causing it to move, the other end of the muscle is attached somewhere pulling on a different body part. That body part will pull back on the muscle so that the overall force acting on the muscle is zero. That other body part is often immovable because of other forces acting on it. It should have enough structural integrity that it does not collapse when the muscle pulls on it (a bone or a joint perhaps). The biceps muscle used to lift your forearm is attached near the shoulder joint. If the shoulder joint were to collapse, you would not be able to lift your forearm (fig. 1.2). We will discuss muscles and their endpoints in more detail later.

Other Forces

In addition to force applied by a muscle to move a body part, there are other forces that arise in the discussion of movement. One is *weight*, or the force of gravity pulling downward on a body part. Another is *friction*, a force that opposes the tendency of two objects in contact to move across

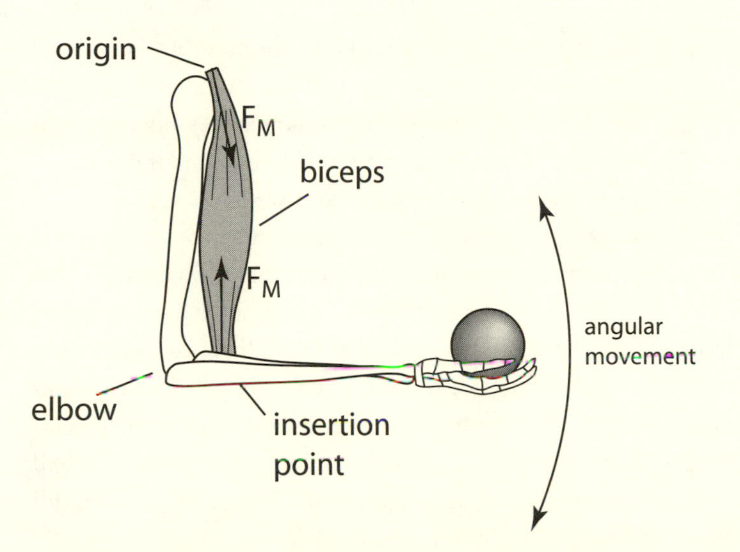

FIGURE 1.2. An example of angular movement. The attachments of the biceps muscle at each end are called the origin and the insertion point. The insertion point moves toward the origin when muscle tension F_M is applied. The origin is fixed and does not move.

each other, and which is particularly important for a person to walk or run. And another force is called the *normal* force, which results when two objects are in contact with each other. And finally there is *tension*, which is a force exerted by a muscle on a body part.

Weight

The weight of any body part is usually involved in its movement. Lifting an arm, holding up your head, and standing on tiptoes are all examples where muscle forces are applied to the body that balance out the weight of the body part. You can hold your forearm level to the ground by tightening your biceps muscle; if you were to relax your biceps, your arm would fall down because gravity pulls it downward. If you are sitting upright, but then begin to dose off, your head may start to fall forward (or backward) if the muscles in your neck have relaxed; gravity is pulling your head downward and your neck muscles must be tense to hold your head up. To stand on your tiptoes, the muscles in the calves of your legs must pull your heels upward off the ground; when you relax those muscles your weight acts to pull your heels (and you) back to level ground.

The weight of an object is proportional to its mass. A bowling ball weighs more than a beach ball because the bowling ball has more mass. The weight of an object can be easily calculated if we know the mass. To do this, we first consider an object in *free-fall*, which means that gravity is the only force acting on the object and it is accelerating downward. This situation works quite well when we consider dense, heavy objects like bowling balls or rocks, and when they fall over short distances; in this situation air resistance can be ignored. It doesn't work so well for very light objects like beach balls or feathers because air resistance can be large relative to the weight of these objects. That doesn't mean that the beach ball and feather don't have weight, it just means that they are not in free-fall when they are dropped.

When an object is in free-fall, it accelerates downward toward the earth with a certain value. Near the surface of the earth that value is about 9.8 meters per second squared, or 9.8 m/s^2. We use the symbol g for this value and call it the acceleration due to gravity. To get the weight of an object we use equation (1) above and write

$$W = mg, \tag{2}$$

where W is the weight of the object, equal to the net force acting on the object when it is in free-fall, m its mass, and g its acceleration when it is in free-fall.

Equation (2) gives us the weight of the object even if it is not in free-fall. The force of gravity still acts on the object so it must be considered when the net force is determined. In other words, an object has weight (just as it has mass) no matter if it is moving or stationary.

Friction

Friction is a force that allows a person to walk or run. Friction occurs when two objects are in contact with each other and they slide (or attempt to slide) across each other. If you put your hands together and move them back and forth, you can feel the force of friction. Each hand exerts a friction force on the other (Newton's third law). If you have glue on your hands, the force may be quite large and it will be difficult to move your hands. If you have liquid soap on your hands, they may be very slippery and you won't feel much friction at all. If you place a book on the table and give the book a quick shove, it may move a short distance across the table but it will likely stop very soon. If you place the book on a slippery surface like ice, then the book may slide for quite a distance because the frictional force is not so large.

You may think of friction as being a force that causes an object to slow down and stop moving. However, friction can also be used to make an object start moving (and also to speed it up). How does a car at rest start moving? If it is on a hill, gravity can act on it to pull it down the hill. But if it is on a flat road (say, stopped at a traffic light), how does it begin to move? In order for the car to start moving, friction must act on the wheels.

The key point here is that friction is a force that opposes the motion of two objects as they slide or attempt to slide across each other. When the driver presses on the accelerator, the mechanical parts of the car try to make the tire rotate. The part of the tire in contact with the road tries to move backward, pushing on the ground as it does so. In response to this action the ground pushes on the tire in the forward direction, causing the car to move forward (fig. 1.3a).

If the car is on ice, then friction is greatly reduced because ice is so slippery. If the driver presses too hard on the accelerator, the tire will start to move and will spin without the car moving forward. The tire moves backward across the ice, but there is very little or no friction from the ground to push the car. To help improve traction, the driver may put sand on the ice or place a thin board under the tire that will increase the friction and might move the car forward. You can also use snow tires or put chains on the tires to help with driving on ice.

Now apply these ideas to walking. Friction between your shoes (or feet) and the ground helps you to walk. It is the relative motion of your shoes pushing backward on the ground that produces a frictional force causing the ground to push you in the forward direction (fig. 1.3b). Just think about the difficulty of standing or walking on ice—without friction, you will likely fall to the ground. Your weight pulls you downward and you may break a bone if the force exerted on you by the ground is large enough.

One other point to make about the force of friction is that it is parallel

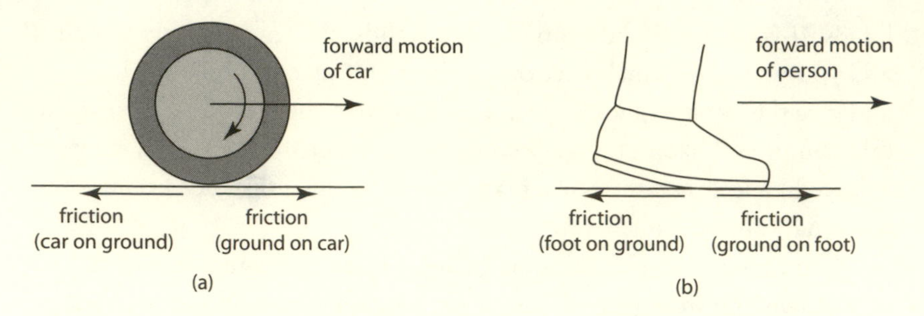

FIGURE 1.3. Action and reaction. As the wheel tries to rotate clockwise, the point in contact with the ground tries to move to the left by pushing on the ground. The reaction force is the ground pushing forward on the wheel and causes the wheel to move forward. A similar situation occurs when walking: pushing backward on the ground causes the ground to push forward on the foot, producing forward motion.

to the surfaces in contact with each other. Thus if a book is sliding to the left across a table, the frictional force is acting to the right, parallel to the book and the table. If the tire of a car is trying to slide to the left on the ground, then the ground exerts a frictional force on the tire that points to the right parallel to their surfaces.

Another idea about friction is that there are two common forms: *static* and *kinetic*. Static friction occurs when the two objects in contact are not actually moving across each other, but are at rest. For kinetic friction, the two objects are moving with respect to each other. For example, suppose you have a large crate that you need to push across the floor; it takes a good deal of force to get it to move, but not as much to keep it moving. Before the crate starts moving, static friction between the floor and the crate acts to oppose the tendency of the crate to move when you push on it. However, after you get it moving, the frictional force that opposes the motion drops to a smaller value. So, once you manage to start the crate moving, don't stop! Less force is needed to keep the crate moving than was needed to start it moving.

Normal Force

The picture of you falling to the ground that I described above brings up another force that may need to be considered: the normal force. Normal does not mean usual or typical or ordinary. It is a mathematical term that means perpendicular. A normal force occurs when two objects are in contact and exert forces on each other that are perpendicular to their surfaces. For example, if you are sitting in your chair right now, then your rear end is in contact with the chair. The chair pushes upward on your rear end and your rear end pushes downward on the chair. If you lean against the back of the chair, it pushes forward on you and you push backward on the support. If you are standing on level ground, the ground pushes upward on you and you push downward on the ground. If a book rests on a table, then the table pushes upward on the book and the book pushes downward on the table.

It gets a bit trickier if you have an object on a sloped surface, such as a book on a table that is lifted up on one end (fig. 1.4). There are several forces acting on the book, but the normal force is exerted on the book by the table in the direction perpendicular to the surface of the book. If the table were to be suddenly removed from the picture, then the book would fall to the ground.

The normal force is often considered to be a support force, or a force that helps hold up an object or keeps it from falling, as for a book on a table, or a person in a chair, or a car on the road. However, normal forces can be used to cause an object to move. If you stand in an elevator, then the floor pushes upward on you to hold you up. If the elevator begins to move upward the increased force exerted on you by the floor will cause you to accelerate upward.

If you were trying to stand on ice, the normal force would be pushing upward on you. But if you were to slip and fall down, the normal force that would arise when you came into contact with the ice would stop

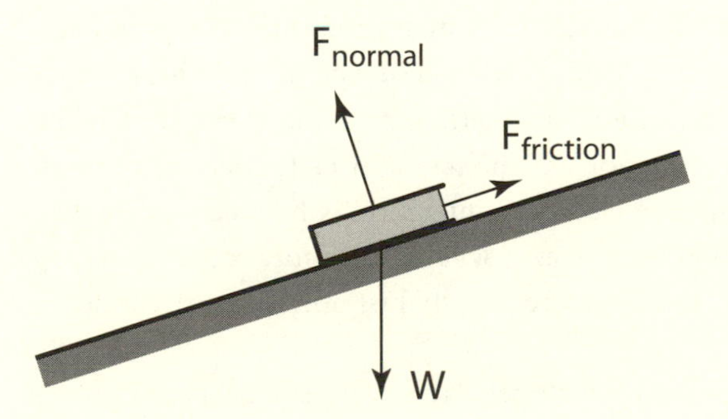

FIGURE 1.4. Forces on a book on a sloped table: the normal force F_{normal}, the frictional force $F_{friction}$, and the weight W.

your fall. This normal force might be large enough to break a bone or cause other damage to your body.

Tension

If you were to play tug-of-war with a friend, then you would both be pulling on the rope between you. The rope would be pulling on both you and your friend with a tension force. Tension in a muscle works in the same way. If you hold your forearm horizontal, there is a tension force in the biceps muscle attached to your forearm and near your shoulder; the muscle pulls on both ends. Suppose that the biceps muscle shown in figure 1.2 exerts a tension force of 200 newtons (N) on your forearm. Because the same biceps muscle also acts on the shoulder, the force applied at that point is 200 N also. Muscles under tension exert the same amount of force at either end.

Often a tension force in a rope is caused by friction or by a normal force. For example, if a rope is attached to an object, say a sling holding up a sprained arm, then the rope is under tension, but it is simply a contact force between the rope and the object. Or in the case of the tug-of-war, the rope in your hands is held tightly so that friction is the main force of interaction between your hands and the rope. When discussing muscles, I will use this idea of tension force being applied to a body part.

What Forces Are Acting?

There are many forces that can act on objects; we have talked about only four: weight, friction, normal, and tension. When considering the motion of an object we will have to look at several different forces acting at the same time, but they will likely each be one of these four. Other forces arise, such as electrical and magnetic forces, but these are usually considered at the microscopic or atomic level. In fact, friction, normal, and tension forces all involve electrical interactions between atoms and molecules at the molecular and atomic level. Other forces that come up are push and pull forces, but these are usually just normal forces between the objects involved.

Think about the book on the sloped table. There are basically three forces acting on the book: the weight of the book pulling straight down, the frictional force that keeps the book from sliding down the slope (and is parallel to the surfaces), and the normal force that pushes perpendicular to the book-table surface (fig. 1.4). If the table were suddenly removed, then the book would fall down to the ground because the net force acting on the book would be the force due to gravity. If the table was very slippery so as to eliminate friction, then the book would slide down along the surface of the table—the normal force would still be there and gravity would be acting, but there would be no friction to hold it in place.

By the way, a tool that helps to determine the motion (or not) of an object is the *free-body diagram*. It is a diagram that shows the forces acting just on that object without including the forces acting on other objects. When the forces acting on one object are isolated, they can be added together (according to rules of vector addition) in order to determine the net force. Vector addition is easy if the forces are parallel to each other; simply add forces that point in the same direction and subtract those that point in opposite directions. In figure 1.1, the forces of the two people pushing on the car add together and the friction force is subtracted. When forces are not parallel, vector addition is much more complicated, and we won't go through the details. However, perhaps you can tell from figure1.4 that, for the net force on the book to be zero, the normal force must be balanced by the part of the weight of the book that points perpendicularly to the table, and the frictional force must be balanced by the part of the weight of the book that points down the slope of the table.

When we talk about movement of the human body, we will try to discuss only those forces that cause the body part to move or to keep it from moving. By isolating the body part, perhaps we can determine whether it will move or be stationary.

Torque

Up to this point we have been discussing forces that act on an object to get it to move. It is also important to discuss where the forces are applied, because the motion of the object could involve rotation rather than long-range motion to the right or left, or up or down (which is called *translational* motion). It is also possible to have situations occur where the net force is zero, but the object moves in a rotational sense rather than undergoing translational motion. These ideas bring us to the topic of *torque*.

While a force is required to change the state of motion of an object, a torque is required to change the *rotational* motion of an object. For an object to begin to rotate or for it to stop rotating, a torque is required. Notice that motion is still involved when an object rotates, so that means that a force is required to change the motion. But in order for the rotational state to change, the force must act in a way that produces a torque. Torque is particularly important for movement of the human body. I have written several times already about using your biceps muscle to lift your forearm. In fact, the entire forearm does not undergo translational motion, but rotational motion—the hand and wrist may move through a significant distance, but the elbow doesn't move at all.

When an object rotates, there is a point or an axis about which the object rotates, called the *pivot point* or the *axis of rotation*. A force can cause the object to rotate if it acts at a point away from the pivot point. For example, if you try to open a door, you usually push on the door knob perpendicularly to the door. However, you could also push near the hinge, but then you would have to push much harder. If you were to push directly on the hinge, you wouldn't be able to get the door to open no matter how hard you pushed.

Not only should the force be applied at a location away from the pivot point, it must act in a certain direction relative to the pivot point for it to produce a torque. If the force acts in a direction that goes through the pivot point, it will not rotate the object. For example suppose you push on the end of the door, but the direction in which you push goes through the hinge. In this case, the door will not rotate. It may wiggle around a bit, but only if the direction of your push force is directed away from the hinge, even by only a tiny bit.

Another simple example of this occurs if you try to balance a long stick on your hand. The point where the stick is in contact with your hand is the pivot point. The weight of the stick can be thought to be concentrated at a point called the *center of gravity* (if the stick has a symmetric shape, then the center of gravity is in the center of the stick). The weight of the stick, acting at the center of gravity, will produce a torque and cause the stick to fall over unless you can keep the center of gravity directly over the pivot point. If the stick begins to fall, then you can quickly shift your hand so that the pivot point is under the center of gravity and the stick will stay in place. Of course this ability to balance a stick requires some agility and quick movements in order to adjust the location of your hand. If you were to stop moving your hand, the stick would fall over.

So the *location* and the *direction* of the applied force are both important parameters in determining the torque caused by a force. We can be more quantitative about torque by defining it as the product of the force and a distance, which we call the *lever arm*. The lever arm is the *perpendicular* distance from the pivot point to the line of the force. The perpendicular distance is used because that has the greatest turning effect. If you have ever used a wrench to tighten a nut on a screw, you know that the best way to do this is to apply a force perpendicular to the wrench. The lever arm is the distance from the screw at one end of the wrench to the other end where you apply the force.

The product of the force and the lever arm is the torque, which can be written mathematically as

$$\tau = Fr_{\perp},$$

where τ is the torque, F is the applied force, and $r_{\perp}$ is the lever arm. The units for torque are the newton-meter and the foot-pound. Perhaps you have looked at the owner's manual of your car and have read where the lug nuts on the wheels should be tightened to so many foot-pounds using a torque wrench.

To see how the expression for torque works, think again about trying to open a door. If you push perpendicularly to the door at the doorknob, then a certain force is required to get the door to move. But if you were to push near the hinge, you would have to push much harder in order to get the same torque. In general, to produce a given torque on an object, a longer lever arm requires less force whereas a shorter lever arm requires more force.

So how can you lift your forearm? The tension in the biceps muscle applies a force to the forearm at a point where the muscle is attached; it is not attached at the elbow, but a short distance from the elbow. If your

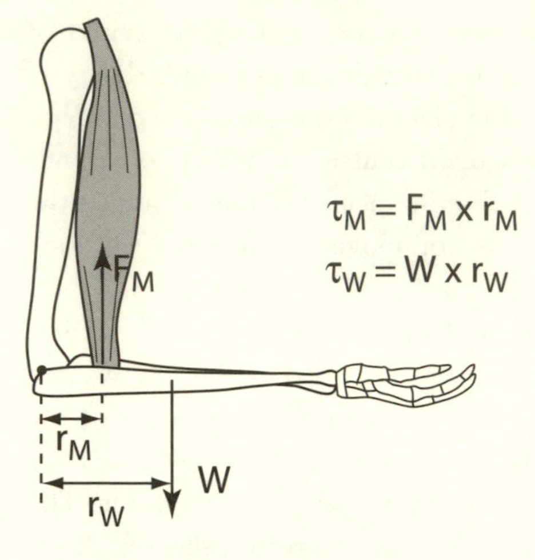

FIGURE 1.5. Rotational motion. When the biceps muscle exerts a force F_M on the forearm, it acts through a distance of r_M measured from the elbow. The resulting torque τ_M causes a counterclockwise rotation that lifts the arm. The weight W of the forearm, acting through a distance of r_W, causes the arm to rotate clockwise when the biceps muscle is relaxed.

forearm is horizontal and your upper arm is vertical, then the biceps is almost vertical as well. The torque produced by the biceps equals the tension in the muscle times the distance from the elbow to where the muscle is attached (fig. 1.5).

Now if you were to relax your biceps muscle then your forearm (not your entire arm) would fall, or rather it would rotate. This happens because the weight of your forearm, which acts at the center of gravity, produces a torque calculated as the weight times the distance from the center of gravity of the forearm to the elbow.

How does the arm start moving if it is initially at rest? For rotational motion we need to consider the net torque. Some torques may cause the object to rotate in one direction, say clockwise, but other torques may cause it to rotate in the other direction or counterclockwise. In order for an object to start rotating from rest, the torque in one direction must be greater than the torque in the opposite direction. Again, consider lifting the arm. If the torque produced by the biceps muscle is greater than the torque produced by the weight of the arm, then the forearm will rotate upward as you lift your arm. If you relax the biceps muscle, the only torque acting on the forearm is caused by gravity, so it will fall back down.

Perhaps you are beginning to see that torque is important in the human body. To lift your arms, to swing your legs when you walk, to stand on your tiptoes, or to bend over and stand back up—all of these involve torque. Note that in some cases the lever arm that exists between a muscle and a pivot point is quite small; this causes the necessary force to be quite large to produce a given torque. In fact, the force may be many times the weight of the object that you are trying to lift.

Torque as a Lever

When discussing torque and its applications to the human body, it is helpful to think of a special type of torque known as a *lever*. A lever is a simple machine that is used to lift heavy objects or to do work; examples include such devices as scissors, a wheelbarrow, and tweezers. When using a lever there are three important considerations. First, there is a location where a force is applied by a person or an object that begins the process of performing the task at hand; this force is called the *effort*. Second, when the lever does its task, it applies a force to the object that is to be cut or lifted or held, and the object exerts a reaction force on the lever at the point of contact; this reaction force is called the *load*. Third, there is a pivot point about which the lever will rotate; this pivot point is often called the *fulcrum*. For the scissors, the pivot point is where the two blades are joined near the center; for the wheelbarrow, the pivot point is where the tire comes in contact with the ground; and for the tweezers, the fulcrum is at the end where the two halves are joined.

There are three types of lever based upon the arrangement of these three points: the first-, the second-, and the third-class lever (fig. 1.6). For a first-class lever, the effort and the load are located on opposite sides of the fulcrum. Scissors are an example of a first-class lever—the effort is applied at the handle and the load is applied at the blades, located on opposite sides of the fulcrum. A seesaw is another example, where one person's weight (either one) provides the effort and the other person's weight is the load. Other examples include a hammer used to pry up a nail, a screwdriver used to pry the lid off a can of paint, and a dolly (or hand truck) used to carry a load of boxes or books. Another first-class lever is a long rod used to lift a heavy object by placing its end beneath the object, allowing it to extend over a pivot, and then applying a downward force on the free end. For tools such as these, the main advantage is that the effort you apply will likely be much smaller than the load because the

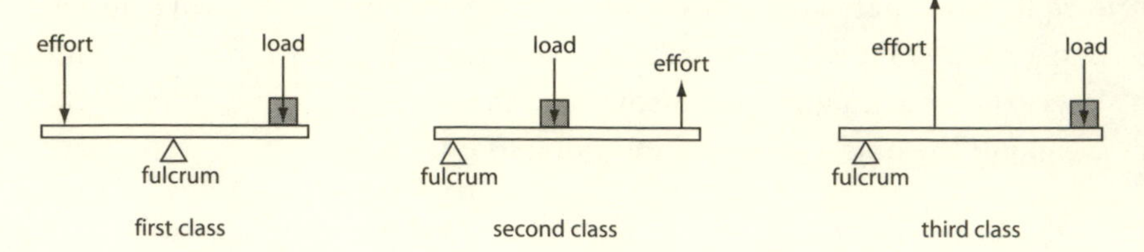

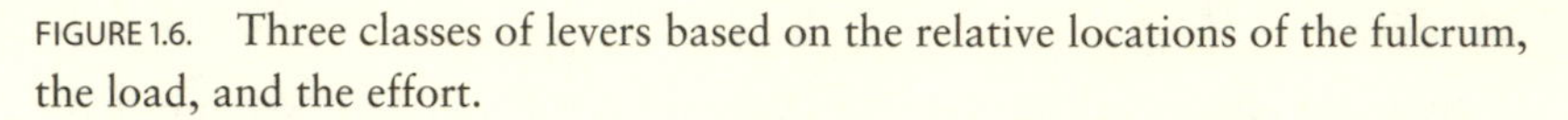

FIGURE 1.6. Three classes of levers based on the relative locations of the fulcrum, the load, and the effort.

lever arm for the effort is quite long compared to the lever arm of the load. This idea is called *mechanical advantage*. In our bodies, there are many situations where the location of a muscle does not provide a mechanical advantage.

For second- and third-class levers, the fulcrum is located near one end of the object and the effort and load are located on the same side. For a second-class lever, the load is located closer to the fulcrum than the effort, and for a third-class lever, the load is located farther from the fulcrum than the effort. (Many of the muscles in our bodies are third-class levers.) A wheelbarrow is an example of a second-class lever—the fulcrum (the tire in contact with the ground) is located near one end, the load is the weight of the material in the wheelbarrow (including the weight of the wheelbarrow itself), and the effort is provided by the person lifting on the handles. Other examples include a nutcracker and a wrench. For the second-class lever, the effort is smaller than the load because the lever arm is longer (mechanical advantage again). Thus, large forces can be applied to accomplish a task, such as lifting the heavy load in a wheelbarrow, cracking a nut, or turning a bolt tightly by applying a much smaller force.

Tweezers represent a third-class lever with the hinge (fulcrum) located at the end where the two blades are joined. The effort is applied somewhere along the middle of the two blades, and the load is applied at the end of the tweezers by whatever is being held (even the other blade if nothing is being held). Other examples include a stapler and a shovel (where one hand at the end is the fulcrum and the other hand lifts the shovel). For the third-class lever, the effort is larger than the load because the lever arm for the effort is shorter than for the load (a mechanical *disadvantage*). Again, this is the situation for many of the muscles in our bodies.

Mechanical Equilibrium

One more item to discuss before we look at the anatomy and physiology of muscles and bones is the topic of *mechanical equilibrium*. If an object is in mechanical equilibrium, then its state of motion remains constant for both translational and rotational motion. This occurs when the net force and the net torque are both zero. If the net force is zero, then all the forces are balanced—just as much force acts in one direction as in the opposite direction. If the net torque is zero, the clockwise and counterclockwise torques are balanced.

When an object in mechanical equilibrium is at rest, it will remain at rest—it will not start moving in either a translational or a rotational sense. This situation is good if we are talking about a bridge because we generally don't want bridges to collapse; bridges are designed and built so that the net force and the net torque on them are zero. If you are holding your forearm horizontal and at rest, then the forces and the torques that act on the forearm must cancel. For a person to stand upright, all the forces and the torques acting on the person must be balanced so there is no net torque. If a person loses his or her balance, then that person may fall because there is a net force or a net torque acting on the person.

Anatomy and Physiology of Movement

In order for your body to move, forces must be applied to various body parts; lifting your arm, nodding your head, and walking or running all require forces. Muscles exert those forces. Forces supplied by bones also play a role, but we'll look first at how muscles work and how they are attached to bones, and then look briefly at bones and joints.

Muscles

There are three types of muscle, all of which are used to move various parts of the body. The main type that we will discuss here is skeletal muscle, which is attached to the skeleton for large external motion of the body. The other two types are used for moving internal parts: cardiac

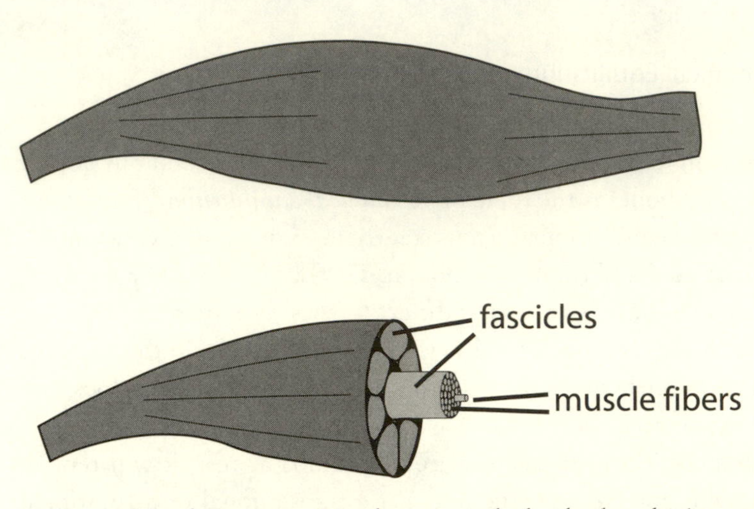

FIGURE 1.7. Muscle structure. Muscles are attached at both ends. A cross-sectional view shows the basic components of muscle tissue: fascicles and muscle fibers.

muscles are used to control the beating of the heart, and smooth muscles are used to move fluids and other materials through the body, such as in the stomach, bladder, and lungs.

Skeletal muscle is considered an organ of the body and consists of muscle fibers, blood vessels, nerves, and connective tissue. Muscle fibers are actually long cylindrical cells up to 100 μm in diameter and 30 cm long depending on location; many fibers are bundled together to form a fascicle; and several fascicles are bundled to form the muscle. Connective tissue surrounds each muscle fiber, each fascicle, and the entire muscle; the connective tissue provides structure and support for the muscle (fig. 1.7).

Each muscle fiber is made up of hundreds to thousands of long, rod-like fibers (myofibrils) that are about 1 to 2 μm in diameter. These myofibrils contain even smaller components or segments (sarcomeres and myofilaments) that expand or contract as needed in order to relax or tense the muscle, even down to the molecular level.

Attachments

Skeletal muscles are attached to bones either directly or by tendons. There are at least two points of attachment, one at each end. One end causes the bone to move when the muscle contracts and is called the *insertion point*. The other end is attached to a bone that does not move (or does not move as much) and is called the *origin*. Either end of the muscle may be directly attached, or it may be attached to bone or other muscles with a ropelike or sheetlike tendon. The muscle usually passes over a joint that serves as the axis of rotation.

Bones and Joints

Bones contribute to the shape and form of our bodies, but they provide several other important functions as well. Some bones provide support for the body, like bones of the legs that hold up the trunk when standing. Some provide protection for the body, like the skull for the brain and the rib cage for the chest cavity. Bones also store minerals and other materials (such as those needed for growth) that are released when the body needs them, they also provide the location for the production of blood cells in the marrow, and, most relevant to this chapter, they are used as levers that move the body when skeletal muscles apply forces to them.

Just as muscles are considered organs of the body, so are bones. They are composed of bone tissue, nerve tissue, cartilage that covers the ends at the joints, connective tissue, and other tissue containing blood vessels. The bone tissue is the primary structural component that provides rigidity for movement of the body.

Most of our discussion has centered on the connection of muscle to bones. But muscles are attached to skeletal cartilages as well. Cartilages are structural components found in locations of the body where flexibility is needed, such as the larynx, the nose, the ear, the rib cage, and the elbow and knee.

Joints are the locations where two or more bones meet. At a joint, the bones may or may not move depending on their purpose. The dominant type of joint (synovial) allows for substantial movement between the bones and is found mainly in the limbs (arms and legs). Synovial joints contain cartilage that covers the ends of the bones, a cavity or capsule that is enclosed by connective tissue, synovial fluid that is very slippery and reduces friction between the bones, and ligaments that connect the bones to each other and keep a joint from moving too much. There are some joints that do not move or have very limited motion; these are usually found in the head, neck, and trunk.[1]

Application of Muscle Tension

When a muscle is relaxed, it is soft and there is almost no tension in it. When the brain tells a muscle to move, it contracts, producing a tension force that pulls on the bone or cartilage to which it is attached. One end of the muscle, the insertion point, is attached to the bone that is expected to move. The other end, the origin, is attached to a bone that will not move or not move as much as the other bone. As the muscle contracts, the insertion point moves toward the origin (fig. 1.2). Using terminology from levers, the tension in the muscle is the effort, and the weight of the object to be moved (the bone plus any other object involved, as when a book is lifted) is the load.

If the contracting muscle causes the body part (bone and object) to move, then it is called an isotonic contraction. But sometimes the object does not move, as when you are trying to lift a very heavy object (a car) or when you are pushing on an immovable object (a wall). The muscle contraction still occurs, but the muscle doesn't shorten; this is called an isometric contraction. During isotonic contractions the muscle will shorten, but there is tension in the muscle whether the object moves or not.

There is also tension in the muscle if the muscle stretches. This happens if you are doing stretching exercises. Before I go running, I usually try to stretch the rear thigh muscles by placing the heel of my foot on the countertop and reaching for my toes. I stretch the calf muscles by pushing against a wall or a tree with my leg and foot extended behind me. I also try to stretch the front thigh muscles by grabbing my foot behind my back and pulling it toward my rear end. If I were to squat, then the front thigh muscles would stretch as they lengthen, but then as I stood up they would contract—in both events the thigh muscles would be under tension.

Types of Movement

There are a number of different ways that a body part can move under the tension of attached muscles. The motion occurs at a joint with the muscle spanning the joint, its origin on one side of the joint attached to the bone that does not move, and its insertion point on the other side of the joint attached to the movable bone. As the bone moves, the insertion point of the muscle moves toward the origin of the same muscle.

The types of movement fall into several categories. The first is *gliding* movement, or translation movement, where two nearly flat bone surfaces slide across each other without significant changes in angle and without rotating. The second is *angular* movement where the angle between the movable and immovable bones increases or decreases; examples include lifting the arm, lifting the leg, bending over, flexing the foot, and bending the head, as well as returning each of these parts to the starting position. The third type of movement is *rotational* movement where the bone twists around its long axis; examples include turning the head or twisting the upper body while the hips don't move.

Other types of movement are very similar to these three categories, but not quite the same. One example is the case where the forearm with the palm of the hand facing upward (as when lifting an object) is rotated to a position with the palm facing downward (as when bouncing a basketball). This example is very similar to rotational movement, but in the forearm there are two bones, the radius and the ulna, and as the forearm rotates downward, the bones cross each other. Other examples include moving the jaw (up and down or forward and backward) and using the hand to grasp.

Classes of Levers in the Human Body

Now let's apply the classes of levers to the human body and look at a few examples. Recall that there are three classes of levers that depend on the relative locations of the fulcrum, the load, and the effort. For the human body, the joint serves as the fulcrum, the weight of the body part is the load, and the force of the muscle is the effort.

In the first-class lever, the load and the effort are on opposite sides of the fulcrum. An example of this class occurs when you hold your head up (fig. 1.8a). A joint at the top of the spine and the base of the skull, called the atlanto-occipital joint, is the fulcrum. The weight of the head is the load. To keep the head from falling forward, muscles in the back of the

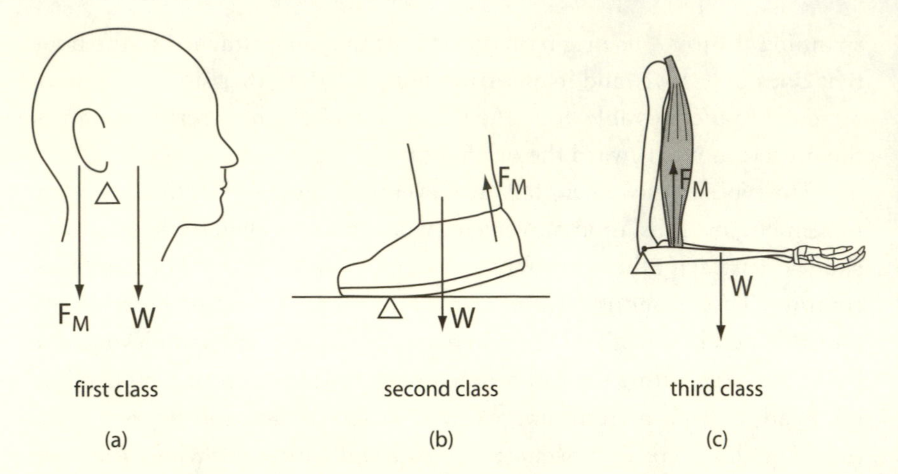

FIGURE 1.8. Examples of three classes of levers in the human body. In each case the weight *W* of the body part is the load and the force applied by the muscle F_M is the effort. The fulcrum, represented by the triangle, is located at a joint.

neck provide the effort; to keep the head from falling backward, muscles in the front of the neck act. The triceps muscle along the back of the upper arm is another example.

For the second-class lever, the fulcrum is near one end and the effort is near the other end with the load located somewhere between them. An important example of this class is the action of standing on your tiptoes (fig. 1.8b). The ball of the foot is the fulcrum and the effort is provided by the gastrocnemius muscle in the calf of the leg. The load, which is nearly your entire body weight, is located between the ball of the foot and the heel. Because the lever arm for the effort is longer than the lever arm for the weight, the effort is less than your weight. This mechanical advantage makes it easy to stand on tiptoes and even to support more weight if you are carrying an object or lifting weights.

For the third-class lever, once again the fulcrum is near one end, but the effort is applied between the fulcrum and the load. Because the lever arm for the effort is shorter than the lever arm for the load, the effort must be greater than the load (a mechanical disadvantage). An example is the biceps muscle located in the front of the upper arm and used when lifting the forearm (fig. 1.8c). Another is the deltoid muscle located at the top of the upper arm, spanning the shoulder joint, used to lift the entire arm. It is difficult to lift very heavy objects because the effort is several times the weight of the object. Muscles used in third-class levers tend to be quite thick and bulky.

Force in the Biceps

In order to get an idea of how much force is required in the biceps muscle, let's look at a numerical example where the forearm is held in mechanical equilibrium. Suppose a person is holding his or her forearm in a horizontal position with the upper arm vertical. The weight of an average person's forearm (and hand) is about 2% of the total body weight. Say the person weighs 160 lb, which means that the weight of the forearm is about 3 lb, and is holding a 10 lb ball in their hand. Because the fulcrum is located at the elbow all of the forces acting on the forearm will be measured from the elbow. The insertion point of the biceps muscle is located 1.5 in. from the elbow, the center of gravity of the arm is located 6 in. from the elbow, and the ball is located 15 in. from the elbow. Assume all the forces are vertical, pointing either upward or downward (fig. 1.9).

Recall that for equilibrium the torques must balance, which means the clockwise and counterclockwise torques must be equal. Notice that the upward tension force in the biceps produces a torque in the counterclockwise direction; the weight of the arm and the weight of the

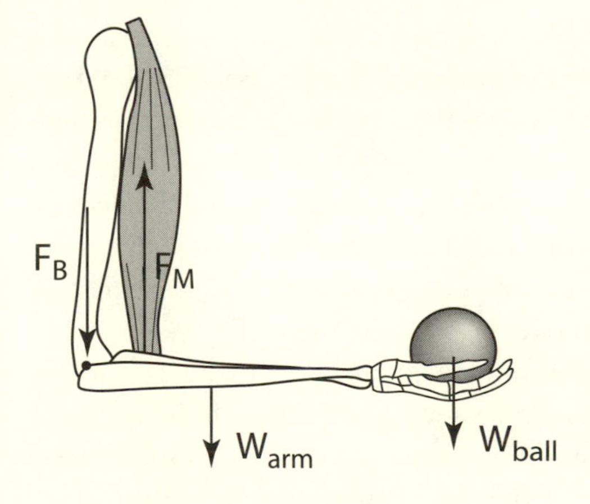

FIGURE 1.9. An example of equilibrium in the forearm. The biceps muscle pulls upward with a force F_M, the weight of the arm W_{arm} pulls downward, and the weight of the ball W_{ball} pushes downward. The force of the bone F_B acting at the joint is necessary to keep the upper arm from collapsing.

ball produce clockwise torques. Thus the solution to this problem is as follows:

$$\tau_{CCW} = \tau_{CW},$$
$$\tau_{biceps} = \tau_{arm} + \tau_{ball},$$
$$F_{biceps}\, r_{biceps} = W_{arm}\, r_{arm} + W_{ball} r_{ball},$$
$$F_{biceps}\,(1.5 \text{ in}) = (3 \text{ lb})(6 \text{ in.}) + (10 \text{ lb.})(15 \text{ in.}),$$
$$F_{biceps} = 112 \text{ lb.}$$

We see that the force in the biceps muscles is quite large—almost forty times the weight of the arm and over ten times the weight of the ball. The main reason it is so big is that the lever arm for the muscle force is quite small compared to the other distances.

Also, you may recall that for mechanical equilibrium the forces must balance. That means that the 112 lb upward force must be balanced by the downward forces. Because the weight of the arm and the weight of the ball add up to only 13 lb, there must be another force of 112 lb – 13 lb = 99 lb acting downward on the arm. This force acts at the elbow and is exerted by the bone in the upper arm.

This example is actually more complicated than just described. For one reason, when the upper arm is held in the vertical position, the force exerted by the biceps muscle is not exactly vertical, but rather is at a slight angle. However, the problem helps to illustrate the large forces that occur in the body, especially in third-class levers.

Force in the Deltoid Muscle

Another example of a third-class lever is the deltoid muscle. Try holding your entire arm straight outward at a right angle from your body so that it is horizontal. With your other hand you can feel the tension in the deltoid muscle near the shoulder. You should also be able to feel the deltoid muscle as it curves slightly downward about 5 to 6 in. from your shoulder; this location is near the insertion point. The deltoid muscle spans the shoulder and is attached just above the shoulder joint.

The curvature of the deltoid muscle is important (fig. 1.10). If the force applied by the deltoid muscle passed directly through the fulcrum, there would be no lever arm and therefore no torque. With the deltoid muscle angled slightly upward from the horizontal arm, the direction

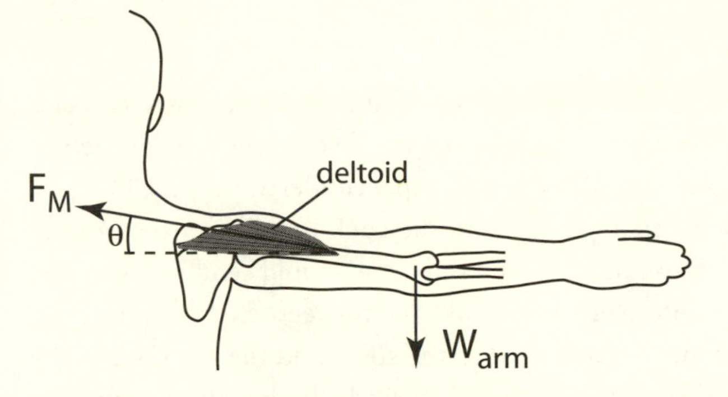

FIGURE 1.10. Torque caused by the deltoid muscle. The muscle is attached to the upper arm and spans the shoulder joint. It makes a small angle relative to the upper arm. The torque produced by the force F_M in the deltoid muscle is used to lift the arm by overcoming the torque caused by the weight W_{arm} of the arm.

of the force does not go through the pivot point, giving a small nonzero lever arm.

Assume that the distance from the shoulder joint to the insertion point is 6 in. and the angle that the deltoid muscle makes is 10°. From these data, we can calculate the lever arm to be $r_{deltoid} = (6\text{ in.})\sin 10° \approx 1$ in. The trigonometric function is needed because, as you may recall, the lever arm is defined as the perpendicular distance from the pivot point to the line of the force.

The weight of the entire arm is about 5% of the total weight of a person, so for our 160 lb person, that would be 8 lb. Also, the center of gravity of the arm is located 10 in. from the pivot point. We apply the condition that the torques must be balanced to find the force exerted by the deltoid muscle:

$$F_{deltoid}\, r_{deltoid} = W_{arm}\, r_{arm},$$
$$F_{deltoid}\,(1\text{ in.}) = (8\text{ lb.})(10\text{ in.}),$$
$$F_{deltoid} = 80\text{ lb.}$$

This value is ten times the weight of the arm. If a 2 lb object were held in the hand located 25 in. away, the force would go up to 130 lb. A 5 lb object would need a force of 205 lb. Clearly the force in the deltoid is quite large compared to the weight of the arm and any other objects being held. Once again, the reason that it is so large is that the lever arm for the muscle force is quite small.

Force on the Spine

Perhaps you have heard someone say that you are not supposed to bend over to lift a heavy object, but you should bend your legs to pick the object up. We can use the idea of torque to help us understand why. In figure 1.11 we see a person bending over with the back parallel to the ground. The weight of the trunk, head, neck, and arms would tend to cause the person to rotate counterclockwise. In order to keep from falling over, there is a bundle of muscles attached to the spine and the hip bones. The origin of these muscles is at the hip bones with the insertion points located along the spine. These muscles are called the erector spinae muscles because they help us to stand erect (and are attached to the spine).

Rather than talk about the forces in each individual muscle, we will model the erector spinae muscles as a single muscle attached at one point on the spine. The pivot point for the upper body is located at the base of the spine at a point called the sacrum. In figure 1.11, the erector spinae

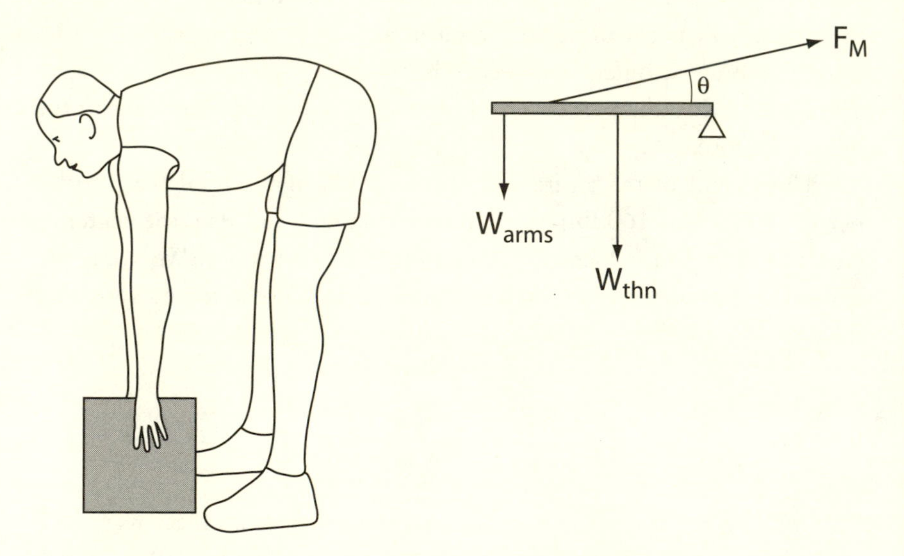

FIGURE 1.11. Another example of torque. For a person bending over to lift an object, a very large force supplied by the erector spinae muscles F_M is required to keep the person from falling over due to the weight of the trunk, head, and neck W_{thn} and the weight of the arms W_{arms}. In this model, the muscle force is assumed to make an angle θ measured from the spine.

muscles will cause a clockwise torque, and the weight of the upper body will cause a counterclockwise torque.

Now you may wonder why an angle is shown between the force in the erector spinae muscles and the spine. As in the case of the deltoid muscle discussed above, if the muscle force were to go directly though the pivot point, there would be no torque. The angle for the deltoid muscle is caused by separation of the deltoid's origin and the shoulder joint. For the erector spinae muscles, it is the curvature of the spine that causes the angle. If you look at a person from the back, the upper portion of the spine bows outward (backward) slightly as you proceed down the back while the lower portion curves inward (forward). This curvature causes the pivot point at the sacrum to be offset by a small distance from the origin of the erector spinae muscles at the hip bones.

Let me give you a few numbers in case you want to verify the calculation. For a person weighing 160 lb, the weight of both arms is about 16 lb and the weight of the trunk, head, and neck is about 90 lb. Suppose that the weight of the arms acts a distance of 2 ft from the sacrum, and the weight of the trunk, head, and neck acts at 0.8 ft from the sacrum. Also, suppose that the force due to the erector spinae muscles acts at 1.3 ft from the sacrum at an angle of 12° from the spine. With these numbers, we find that the erector spinae muscles exert a force of about 385 lb on the spine, which is 2.4 times the weight of the person. If the person were picking up an object that weighs 50 lb, then the force would be about 750 lb, or almost five times the person's weight!

In addition to the forces in the muscles, there are also large forces acting on the sacrum (or the lower part of the spine). Perhaps you can see why we might strain a muscle in the back, or cause injury to the spine (a slipped or ruptured disk) if we aren't careful. If we consider the person squatting down to pick up the object, then the majority of the force is in the legs. There is still some force in the back, but it is much less. It is the act of leaning over that causes large forces on the muscles and on the spine.[2,3]

Balance

When an object is in mechanical equilibrium, the forces and torques are balanced. When gravity is one of the forces acting on the object, then equilibrium is either *stable* or *unstable*. If the object is in stable equilibrium, it will remain in its current position or will return to that position even if it is tipped slightly. An example would be a box sitting on the floor. If the box could be balanced on an edge, it could still be in mechanical equilibrium but it would be unstable; if it were moved very slightly, it would fall from its position.

Stable equilibrium with gravity present occurs if the object has a base that is large enough to support it without tipping over. The box just mentioned may have a wide base, but even a smaller base can work. Take a meter stick for example. Its most stable position is flat on its side on a table or the ground. However, you can get it to stay on its edge if it is flat. And if you are very careful you can get it to stand on one end in an upright position if the end is also flat. It would not be very stable, but for a very, very small movement, it would return to its upright position. It would, however, be very easy to knock it over.

The key to stable equilibrium is that the center of gravity of the object must be over a large enough base. This condition is true for people as well. When you are standing, if your center of gravity is over the base formed by your feet, you will be in mechanical equilibrium (fig. 1.12). If you spread your feet apart then you have a wider base and will be more stable. If you pull your feet together then you can still be stable, but less so.

This works even if you stand on one foot, but you have to shift your

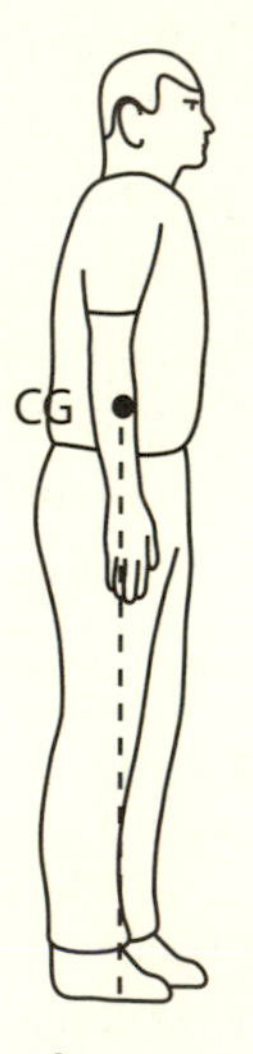

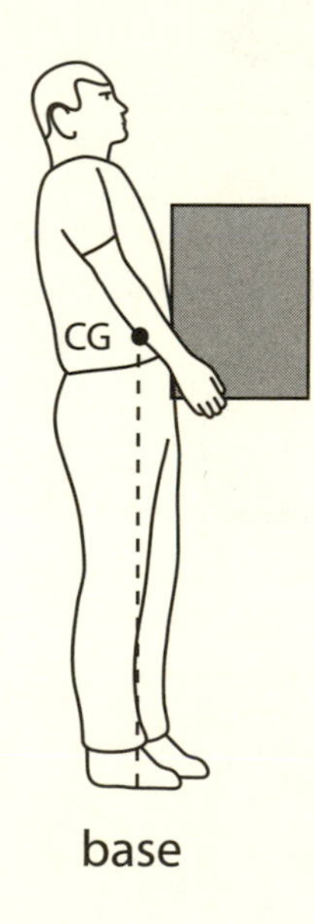

FIGURE 1.12. Stable equilibrium. A person's center of gravity (CG) must be located over the base (the feet) for balance. The person must shift his or her body, causing a change in the center of gravity, when holding a heavy object, standing on one foot, or bending over.

body so that your center of gravity is over that foot. To do this, you would thrust your hips in one direction and your shoulders in the other direction, changing the shape of your body and changing the location of your center of gravity. This position is fairly unstable, but you sense your balance and adjust your position so that you keep the center of gravity over your foot.

The change in center of gravity of a person carrying a heavy object must be taken into account by repositioning. If the object is carried in front, as when a woman is pregnant, she leans backward with her head and shoulders farther back than normal. If the object is carried on the side, such as a backpack slung over one shoulder, the person leans in the opposite direction to maintain balance. If the backpack is on the back, the person leans forward slightly.

Walking

Part of the ability to walk has to do with the idea that the person is out of balance for brief periods of time. To start walking, a person swings one leg forward, causing the center of gravity to be beyond the base of the stationary foot (past the toes). This action causes the person to lean forward while being out of balance for a moment until the front foot comes in contact with the ground. In forward motion, momentum carries the person through the upright position so that when weight is transferred to the front leg, the other leg can swing forward to take the next step. The process continues as the person walks forward.

You may recall what I mentioned earlier: in walking, the foot pushes backward on the ground (a frictional force), and the ground pushes forward on the foot (Newton's third law). It is the combination of swinging one leg forward and pushing backward on the ground with the other leg that moves a person forward. But as the forward leg is placed on the ground, there is a frictional force acting to slow the person down slightly. With the forward leg extended in front of the center of gravity, and the other leg coming up off the ground, the person is unbalanced briefly as well. Both of these actions cause a slight slowing down, but again the forward momentum continues to carry the walking person forward.

The friction may surprise you or seem counterintuitive, but just imagine trying to walk on ice. If you were to try to take a quick step forward, your rear foot would likely slip backward and you might fall. Or if you were to be successful at landing on your forward foot, it might slip forward, causing you to fall.

Difficulty of Movement

One final topic in this chapter has to do with problems that may restrict motion of the body. These problems arise in the muscles, the bones, the joints, and/or the nervous system (including the brain). If an injury or sickness occurs there may be damage to one or more of these systems. Problems may not actually restrict motion, but may cause pain such that the person does not want to move the affected area. There are entire fields of medicine devoted to treating such conditions that arise.

Muscles may sustain injury by being cut or by overuse. Athletes are subjected to large forces that can tear a muscle or a tendon. Underuse will weaken muscles, and no use may cause muscles to atrophy or waste away.

Bones will break if very large forces act on them, particularly sideways forces. Someone who falls from a high position may break many bones. As a person ages the density of bone material decreases, which may lead to osteoporosis. Hip fractures in the elderly are common because the forces on the bony part of the hip joint are quite large.

Problems with joints may be the worst of all because many of them arise even in the absence of injury and because they affect two bones. Arthritis is the most common disease of the joints. Osteoarthritis is a degenerative condition caused by long-term use, so it is mostly seen in the elderly. Rheumatoid arthritis and gout are both inflammatory conditions. Bursitis and tendonitis are caused by injury or overuse.

Finally, certain areas of the brain control the movements of the skeletal muscles. If the brain and/or the nervous system are damaged, either by physical injury or chemically, then proper electrical stimulation may not occur even though there is no problem with the muscles.

Summary

I have just touched on some of the basic ideas of motion of the human body. Muscles exert forces on various body parts to get them to move or to keep them from moving. Other forces play important roles as well, including weight and friction. Joints serve as places where rotation occurs. This topic is much more complex than I have just described. Still, these basic physics concepts should help you to understand a few things about motion, both translational and rotational, and the forces involved that cause the body to move.

TWO

Fluids and Pressure

How does blood flow? What happens when you cough? What is glaucoma? Why do older people's bones break easily? How is a knife able to cut through tissue? Each of these questions has something to do with *pressure* in the body—in the circulatory system, in the lungs, in the eyeball, on a bone, and on tissue. In all of these examples, fluids (both liquids and gases) or solids exert forces on the body, either internally or externally.

Instead of focusing on force, it is often better to discuss pressure. Pressure is simply defined as force applied over an area, and is particularly useful for describing the effect of force on fluids. In this chapter we will look at some basic physics properties of fluids and describe in more detail a number of applications of pressure in the body.

Pressure

Swing a baseball bat to hit a ball, and the force that the bat applies to the ball causes it to move in a different direction. Throw a baseball with your hand and the ball will move (we hope) in the direction you want it to go. Both of these examples can be described using basics laws of physics, such as Newton's laws or kinematics. But when you swing the bat in the air, where does the air go? When you swish your hand in the air, how does it make the air move?

The important quantity is pressure. To move a fluid, either a liquid or a gas, the force must be applied over an area. If the force acts at a point in the fluid, the fluid usually just moves out of the way; this situation applies to the baseball bat moving through the air. But if the force is applied over an area, large portions of the fluid can move. For example, when you fan yourself with your hand or when you move your hands to swim, you are moving significant amounts of air or water in the direction your hand is moving. The force is applied by your hand over an area, rather than at a point, which causes the fluid to move. Thus, pressure p is defined mathematically as force F divided by area A, written as

$$p = \frac{F}{A}.$$

There are many different units for pressure. You are probably most familiar with the pressure in the tires of a car or a bicycle, which is measured in pounds per square inch, or psi, lb/in.^{2}. Another common unit is inches of mercury (in.-Hg), or torr; meteorologists usually report barometric pressure around 30 in.-Hg (which is about 15 psi). Blood pressure is measured in millimeters of mercury (mm-Hg). Other units include the atmosphere (atm), the pascal (Pa), the kilopascal (1 kPa = 1000 Pa), the hectopascal (= 100 Pa), the bar (= 100,000 Pa), and centimeters of water (cm-H_2O).

Pressure in a fluid has three possible sources. One source is the weight of the fluid itself; if you swim to the bottom of a swimming pool and feel the pressure on your ears, then you experience pressure due to the weight of the water above you. Another source of pressure is an external force applied to the fluid; when a water gun is squeezed, the pressure builds up enough that water is forced out of the gun. The third is the internal forces due to interaction of the molecules of the fluid, particularly in a gas; the pressure in the tires of a car illustrates this kind of pressure. All of these types of pressure occur in the human body, so we will look at them in more detail.

Pressure due to Weight of a Fluid

The weight of a fluid produces pressure below its surface. If you dive to the bottom of a swimming pool or fly in an airplane, you will notice a change in pressure on your ears. Also, you may notice that you have to submerge yourself only a short distance (a few feet) in water to feel the pressure, but you have to change elevation in the air by hundreds of feet

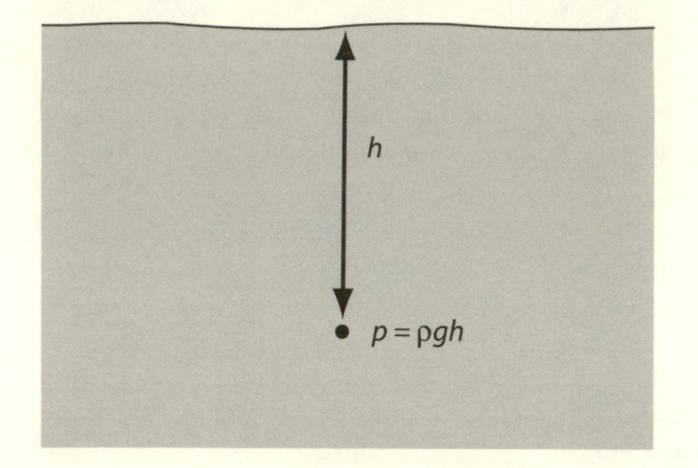

FIGURE 2.1. Pressure p due to the weight of a fluid at a depth h below its surface. The density of the fluid is given by ρ and the acceleration due to gravity is g.

to feel the effect. The difference in type of fluid plays a role in the cause of pressure and is expressed through the density of the fluid. The pressure p due to the weight of a fluid can be calculated using the equation

$$p = \rho g h, \tag{1}$$

where ρ is the density of the fluid, g is the acceleration due to gravity, and h is the vertical depth (not height) beneath the surface of the fluid (fig. 2.1). This expression assumes that the density of the fluid is constant, which is quite accurate for liquids, but not so for gases. However, it is a good approximation for gases if h is small, like a few feet or so.

A similar expression arises when the difference in pressure between two points beneath the surface of the fluid is considered. If the pressure at one depth is p_1 and at another depth is p_2, then

$$\Delta p = p_2 - p_1 = \rho g \Delta h, \tag{2}$$

where Δh represents the difference in depth of the two locations. This expression helps us understand that, as the depth increases, the pressure increases as well. In other words, if point 2 is located below point 1, then the pressure at point 2 is greater than the pressure at point 1. Thus, at higher altitudes above the surface of the earth, the pressure is less than at ground level. Similarly, below the surface of water, the pressure at greater and greater depths is larger than the pressure at sea level.

Let's look at a couple of examples. Atmospheric pressure at ground level (or sea level) is approximately 30 in.-Hg, or about 1 atm. In Denver, which has an altitude of about 1 mile above sea level, the atmospheric pressure is roughly 25 in.-Hg. By contrast, humans can dive to about 100 ft or so below the surface of water where the pressure is about 3 atm. Any lower and a submarine would have to be used. At depths of 1 mile below the water, the pressure is more than 150 atm!

Clearly the weight of a fluid is an important contributor to pressure. Even in the human body, blood pressure can vary depending on whether a person is standing or lying down. Blood pressure will be discussed in more detail later.

Pressure due to an External Force

Suppose you take a balloon that has been inflated. If you squeeze the balloon with your hands or sit on it, you increase the pressure in the balloon. If you are able to press hard enough, the balloon may burst. This external force that you apply to the balloon increases the pressure of the air in the balloon. Other examples include squeezing a water gun, or pushing the plunger on a syringe, or operating a hydraulic jack.

The idea described here can be explained using *Pascal's principle*. This principle states that, when pressure is applied to a fluid in an enclosed container, the pressure is transmitted throughout the fluid, even to the walls of the container that holds the fluid. So, when we squeeze a balloon, the pressure of the air in the balloon increases and is applied to the inside surface of the balloon. If the pressure (or force) is large enough, it can rupture the balloon. If a force is applied to the plunger of a syringe, the pressure in the liquid increases and can overcome the surface tension that keeps the liquid in the syringe, so that it can come out through the needle.

There are many examples of Pascal's principle in the human body. During pregnancy, when a baby is large enough to press on the mother's bladder, she has to go to the bathroom often. Feeling a person's heartbeat is another example; when the heart contracts, it presses harder on the blood in the circulatory system; this increased pressure is transmitted through the blood to the walls of the arteries, which expand slightly; then the heart relaxes and the walls of the arteries relax slightly as well. We will discuss these two examples in more detail later.

Pascal's principle can be used to describe effects due to the atmosphere when making pressure measurements. If you were to measure the pressure in the tires of your car, you would use a tire gauge. The pressure reading should indicate a value of about 35 psi if the tire is properly inflated. However, if the tire is flat, the gauge will read 0 psi. There is still some air in the tire, but it will be at atmospheric pressure, which is about 15 psi.

To handle this discrepancy, we need to be careful. There are three different pressures that need to be defined. First, the gauge pressure p_{gauge} is the pressure that we would measure with a typical gauge, such as a tire gauge or a blood pressure gauge. Second, the atmospheric pressure p_{atm} is the pressure of the air in the atmosphere, typically at ground level or sea level. Third, we have the absolute pressure $p_{absolute}$, which is defined as the sum of the gauge pressure and the atmospheric pressure. Mathematically, we write the relationship as

$$p_{absolute} = p_{gauge} + p_{atm}. \qquad (3)$$

Thus, we have to be careful when reporting pressure measurements to distinguish between gauge pressure and absolute pressure. Most pressure measurements in the human body are reported in gauge pressure, but we may need to use the absolute pressure in some applications. The tire pressure mentioned above represents a gauge pressure.

Pressure due to Interaction of Molecules

The third cause of pressure in a fluid is the interaction of molecules of the fluid with each other and with objects in contact with the fluid. This occurs mainly in gases, although some of the concepts that arise can be applied to liquids as well. For example, the pressure of air in a tire is caused by the molecules of gas colliding with each other and with the walls of the tire. When a tire gauge is used to measure the pressure, the gas in the tire comes into contact with the gauge, pushing on it. If more air is added to the tire, there are more molecules interacting with each other and the pressure increases.

In some cases, when more air is added to an enclosed container, the container expands so that the pressure remains constant. For example, suppose you have a crumpled paper bag. If you begin to blow it up, it expands as more and more air enters the bag. The pressure remains fairly constant until the bag is fully expanded. At that point, if more air goes into the bag, the pressure goes up quickly. If you close off the end of the bag and give it a hard slap, the external pressure supplied by your hand causes the pressure to increase very rapidly and the bag pops—this is Pascal's principle in action!

This type of pressure is the main consideration used in various gas laws, such as the *ideal gas law* and *Boyle's law*. The ideal gas law relates the absolute pressure p of gas in a container, the volume V of the container, the number of molecules N or moles n of gas, and the temperature T of the gas in the formulas

$$pV = Nk_BT \tag{4a}$$

and

$$pV = nRT, \tag{4b}$$

where k_B is Boltzmann's constant, which has a value of 1.38×10^{-23} J/K, and R is the universal gas constant, which is 8.31 J/(mol K) or 0.0821 L atm/(mol K). Equation (4a) is the microscopic form of the ideal gas law that deals with molecules at the molecular level, and equation (4b) is the macroscopic form that deals with the number of moles of gas. This gas law can be used to determine how much air is in the lungs when one inhales and exhales.

Boyle's law is a special case of the ideal gas law. If the number of molecules and the temperature of the gas remain constant, then the product of pressure and volume is constant, or

$$pV = \text{constant}. \tag{5}$$

This equation says that the pressure and volume of a gas are inversely related. In other words, if the volume of an enclosed container were to increase, the pressure would decrease. Because gases fill the container that they are in, if the volume of the container is large, the molecules of the gas are relatively far apart and don't collide with each other or the walls of the container as much, resulting in a lower pressure. If the volume of this gas were to decrease, the molecules would be closer together and collide much more often, resulting in a higher pressure. Breathing is a prime example of Boyle's law in action. To inhale, the chest cavity increases in volume, causing the pressure to decrease. Air rushes into the lungs from outside the body through the nose and/or mouth. To exhale, the chest cavity decreases in volume, causing the pressure to increase, and air rushes out. We will explore this topic in more detail later.

Even though Boyle's law is applied to gases, it can be used qualitatively to describe liquids as well. If pressure from an external source is applied to an enclosed liquid, the pressure will increase, and the container will decrease slightly in size. An example is the cycle of the heart: when the heart contracts, the blood pressure increases, and when it relaxes, the blood pressure decreases.

Pressure in Solids

We have been discussing pressure in fluids, but pressure also occurs in solids. Physicists often use the term *stress* instead of pressure when discussing solids. Some of the same causes of pressure occur for solids, such as the weight of a solid object or application of an external force.

Suppose a woman weighing 110 lbs puts all of her weight on the small area of one high-heeled shoe, say 0.5 $in.^2$; then the pressure is 110 lbs/0.5 $in.^2$ or 220 psi, which is about 15 atm. However, if a 300 lb football player puts all of his weight on one heel of his flat shoes with an area of 10 $in.^2$, the pressure is 30 psi, which is about 2 atm. The smaller woman will be much more likely to damage the floor than the large football player!

Bones and joints in the body have pressures exerted on them. In the average person standing on one leg, about 94% of his or her weight acts on the knee joint. (About 6% of the weight is below the knee joint.) Runners often complain of problems with their knees, which arise not only from their weight, but also from the additional force required to change the direction of the leg, from moving downward to moving upward. Bones break when the pressure is too large.

When subjected to pressure (or force in general) solids flex or deform by an amount based on their elastic properties. Elastic properties arise because the atoms of the solid material are bonded to other atoms and tend to act like springs. As a force is applied, these "springs" will stretch or compress, which causes the material to change shape.

Pressure in a solid is usually referred to as stress. The word "stress" in health care has different meanings, but to a physicist it is the force applied to the object divided by the area over which it acts—thus it is almost identical to pressure. The stress causes a deformation, called *strain*, in the solid material. If the deformation is elastic, then when the stress is removed, the object returns to its original shape. If the stress is too large, then the strain will exceed the *elastic limit* and the object will be permanently deformed. The strain could even exceed the breaking point and result in a fracture.

A couple of examples are in order. When we walk on the floor, our shoes push on the floor and actually cause the floor to flex a bit. When we step on a spot, the floor flexes and then returns to its original position after we step away from the spot. However, suppose we were to drop a sharp object, such as a knife with the blade tip pointing downward on the spot. The stress applied on the floor would likely be enough to cause the strain on the floor to exceed the elastic limit, resulting in a break or tear.

A similar situation occurs for the human body. Suppose you jump up from the floor and land on your feet. The bones in your foot will experience a large amount of stress when you land. Usually everything is just fine and you go on your way with no problems. But suppose you jump from a point 10 ft above the floor, again landing on your feet. In this case, the stress applied to the bones in your feet may be significant enough to break a bone or two—the elastic limit has been exceeded, as well as the fracture point.

These concepts are used quite often by construction engineers as they design bridges and buildings. Parameters that are used in this area are *Young's modulus*, *bulk modulus*, and *shear modulus*. All of these parameters quantify how much deformation occurs in a solid when pressure is applied to it.

The rest of this chapter is devoted almost exclusively to fluids.

Fluids in Motion

The ideas of pressure that we have discussed apply when fluids are at rest and also when they are moving. But there are certain characteristics of moving fluids that we need to explore. The area of physics that deals with fluids in motion is called *fluid dynamics*. It is a very complicated field of study, but there are several concepts we can use to describe a number of applications in the human body. Let's look more closely at three areas: characteristics of fluid flow, flow rate, and viscosity.

Characteristics of Fluid Flow

Fluids often exhibit a number of very complicated behaviors, but in many instances their behavior is less complicated and can be described using relatively simple mathematical models. We will look at several of these characteristics and see how they apply to fluids that flow in the human body.

One characteristic has to do with the flow of a fluid at a specific location—whether it is *steady* or *unsteady*. In steady flow, when a fluid flows along a certain path (for example, through a tube or pipe), its velocity at a particular point does not change. Suppose you measure the speed of water flowing in a river at a certain location. If you were to return a while later, you would expect the speed of the water at that same location to be the same as it was when you measured it previously, provided nothing has happened in the meantime, like a deluge of rain or a boat passing by. Even if someone were to throw a rock into the river, the flow would change slightly, perhaps for a few moments, but then it would likely continue as before. Another example is blood flow through a certain part in the body; it will have the same speed provided there are no changes in conditions. However, if the person's level of activity changes then there could be significant changes to the speed. Even as the heart beats there are changes in the speed of the blood, but one would expect that the speed of blood flowing in, for example, the aorta would have the same speed a fraction of a second after contraction of the heart each time the heart contracts.

Another feature of fluid flow has to do with whether it is *rotational* or *irrotational*. Tornadoes and hurricanes exhibit rotational behavior, as does a whirlpool observed when the water runs out of a bathtub. Small whirlpools (or eddies) even arise when you swish your hands in the water.

Another characteristic of a fluid is whether it is *viscous* or *nonviscous*. In a viscous fluid there is friction between the molecules of the fluid, or between the fluid and the tube or pipe in which it is moving. A viscous fluid that is moving requires energy to continue moving. A nonviscous fluid can flow without losing energy. Gases tend to be much less viscous than liquids; water is less viscous than motor oil. There are only a few fluids that are nonviscous: they are *superfluids* such as liquid helium-3 and liquid helium-4, which can exist only at temperatures close to absolute zero. Not only do superfluids have no viscosity, they also have other characteristics, such as infinite thermal conductivity and very high order (zero entropy), which are outside the scope of this book.

The final characteristic to mention is whether a fluid is *compressible* or *incompressible*. When pressure is applied to an incompressible fluid its volume does not change. An incompressible fluid has a constant density. Liquids are closer to being incompressible than are gases because gases can be easily compressed or expanded in volume.

For real fluids, we often talk about two types of flow: *laminar* and *turbulent*. Laminar flow is often called streamline flow and usually applies when the fluid is flowing at low speeds (and also at higher viscosity and in a smaller region). For a viscous fluid in laminar flow, the fluid near the walls of the tube moves more slowly than the central part of the fluid; this is due to friction between molecules of the fluid and the tube or pipe in which it is flowing. Turbulent flow is very different from laminar flow because the fluid flows in irregular patterns, its velocity changing often, with little if any predictability. At higher speeds (and lower viscosity and larger region of flow), fluids are more susceptible to turbulent flow.

Both types of flow are observed in the human body. Most blood flow is laminar, particularly when the speed of the blood is small, as in smaller arteries and capillaries. Laminar flow also occurs in the larger veins, but turbulent flow happens as well. Turbulence usually occurs when a sudden change in pressure causes the blood to speed up rapidly. A couple of examples may help to illustrate this point. When the heart contracts there is a sudden increase in pressure that causes blood to push open one of the valves (say, the aortic valve), which results in blood rushing quickly into the next region (say, the aorta). This sudden flow of blood becomes turbulent and makes a distinctive sound that can be heard with a stethoscope. Another example occurs when blood pressure is measured—when the pressure of the cuff on the arm decreases and reaches a certain critical point, there is a rapid flow of blood through the artery that results in turbulence, and a sound can be heard with a stethoscope.

Flow Rate

Think of a fluid flowing through a tube, through a pipe, or along a channel. The flow rate Q of the fluid is defined as the volume ΔV of fluid flowing past a particular location divided by the time Δt it takes for that amount of fluid to flow. Mathematically, we write the flow rate as

$$Q = \frac{\Delta V}{\Delta t}. \tag{6}$$

For example, if it takes five minutes to pump 15 gallons of gasoline into the tank of your car, the flow rate is 3 gal/min.

When this definition is applied specifically to a liquid that is moving through a tube, pipe, or channel (not a gas), the flow rate is constant throughout as long as there is no liquid added or removed. This result is observed because a liquid is (nearly) incompressible: a certain volume of liquid flows per unit time through the tube, pipe, or channel. Consider the gasoline example above: the gasoline flowing through the hose from the pump has a flow rate of 3 gal/min and the tank is filling up at the rate of 3 gal/min. We will return to this concept in just a moment.

As it turns out, flow rate can be determined in another way, by considering the cross-sectional area A of the tube, pipe, or channel and the speed or velocity v with which the fluid is moving past that section of the tube. It is calculated using the formula

$$Q = Av. \tag{7}$$

Specifically for a liquid (again, because it is incompressible), where the flow rate is constant in a tube, pipe, or channel, we can use equation (7) to determine what happens to a liquid moving through a tube whose size changes. If the cross-sectional area changes, the speed of the fluid must change in such a way that the flow rate remains constant. That is, the flow rate at two different locations in the tube, points 1 and 2, must be the same, or $Q_1 = Q_2$. Thus, if the area changes from A_1 to A_2, the speed must change from v_1 to v_2, such that

$$A_1v_1 = A_2v_2. \tag{8}$$

This equation is often called the *flow-rate equation* (fig. 2.2). It says that, if the area of the tube, pipe, or channel increases, the speed of the

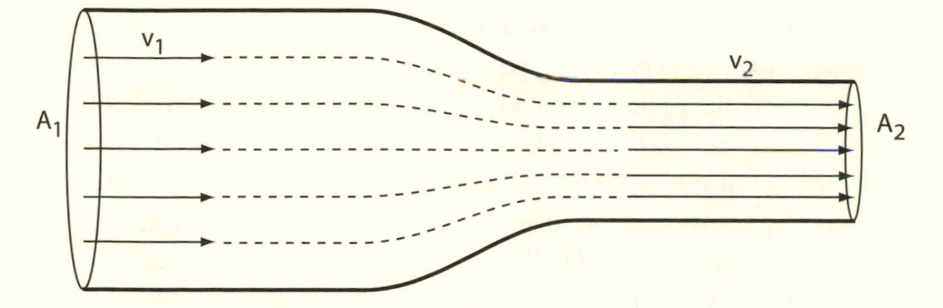

FIGURE 2.2. Flow through a tube with different cross-sectional areas A_1 and A_2. Fluid flows faster (velocity v_2) in the smaller section of the tube than in the larger section (velocity v_1).

fluid must decrease in the larger section of the container. For example, if you use a hose to water the flowers, in order to reach the back of the flower bed you can place your thumb over the end of the hose, which causes the water to squirt farther; the smaller area of the nozzle results in a larger speed. The flow-rate equation also applies when a bucket is filled with water; the surface of the water in the bucket moves upward at a much slower speed than the water coming out of the faucet because the cross-sectional area of the water in the bucket is much larger. If a bucket full of water has a hole in it, the speed of the water coming out of the hole is greater than the speed of the surface of the water as it falls. Another example is that of a river as it flows along at a certain speed at a certain location; if the river were to become narrower at another location along its path, the speed of the water would be greater; if the river were to become much wider at still another location, then the water would slow down quite a lot.

Viscosity

Viscosity is defined in terms of friction between the molecules of a fluid, as well as between the fluid and the walls of the tube or pipe in which it is moving. Viscosity is an *internal resistance* of the fluid to flow. Fluids that have high viscosity do not flow easily; the lower the viscosity, the easier it is for the fluid to flow. Gases have much lower viscosity than liquids, so that gases flow quite easily. Oil and grease are liquids that have high viscosities, but they are used as lubricants; they have a special ability to remain in place between, say, two pieces of metal so that there is reduced friction between the metal pieces.

The viscosity of fluids depends on temperature. If you store molasses in the refrigerator, it may come out of the container rather slowly when you try to pour it on your pancakes. However, if you heat it up in the microwave for a few moments, it will pour much more easily. Motor oil also has this characteristic: oil with higher viscosity is used in the summer and oil with lower viscosity in the winter. Multigrade oils, such as 10W-40, help to reduce the temperature dependence of the viscosity so that they can be used year round. They are designed with different properties to maintain their viscosity at higher temperatures (so that they continue to lubricate the metal parts of an engine), while not being too thick at cold temperatures (making the engine hard to start).

Because viscosity is a resistance to fluid flow, a push or force must act on a viscous fluid in order for it to flow. Gravitational pull can make a viscous fluid flow "downhill," such as water flowing down a stream or river. Buoyant forces can cause fluids that have a lower density than their surroundings to flow upward, as when warm water near the bottom of a pot rises to mix with cooler water (convection). For a viscous fluid to continue to flow through a horizontal pipe or tube, pressure must be constantly applied. A nonviscous fluid flowing through a horizontal pipe or tube does not need a constant application of pressure to keep it flowing; once it starts moving, it continues to move. The greater the pressure applied to a viscous fluid, the greater the flow of the fluid. If the pressure is removed, then the viscous fluid stops flowing.

Let's return to the idea of flow rate and see how viscosity affects it. One model shows that the flow rate Q of a fluid through a tube or pipe depends on several factors: the pressure difference Δp along the tube, the radius r and length L of the tube, and the viscosity η of the fluid flowing through the tube (see fig. 2.3). Mathematically, this model is expressed as

$$Q = \frac{\pi r^4 \Delta p}{8 \eta L}, \tag{9}$$

which is called Poiseuille's law. From this model we see that the greater the pressure difference along the tube or pipe, the greater will be the flow

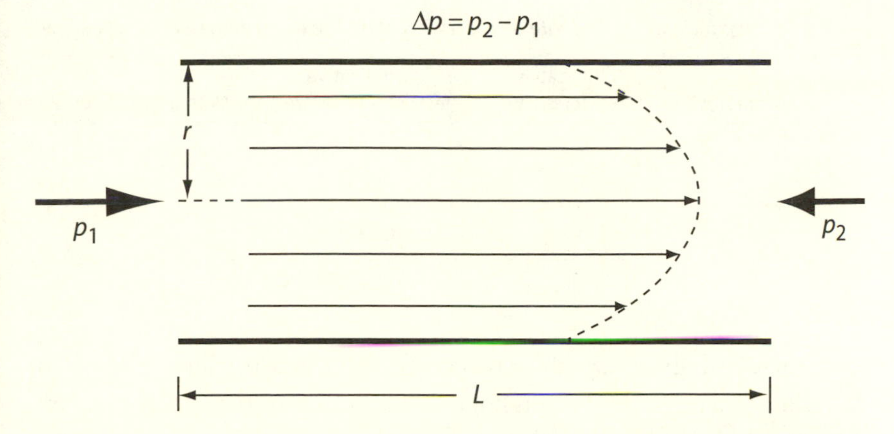

FIGURE 2.3. Flow of a viscous fluid through a tube given by Poiseuille's law. The flow depends on the length L of the tube, the radius r of the tube, the pressure difference Δp across the ends of the tube, and the viscosity of the fluid flowing through the tube. The pressure difference is found as the difference between the pressures p_1 and p_2 at the two ends of the tube.

rate. The flow rate is inversely proportional to the viscosity, so the lower the viscosity, the greater the flow rate.

The flow rate is inversely proportional to the length of the tube, so that the longer the tube, the smaller the flow rate, and vice versa. As an example, consider water from an outdoor faucet that is flowing quite rapidly. If a very long garden hose is attached to the faucet, the flow will decrease.

Poiseuille's law also shows that the flow rate is most strongly dependent on the radius of the tube—it is proportional to the fourth power of the radius. This means that, if the radius were to double, with all other factors remaining the same, then the flow rate would increase by a factor of $2^4 = 16$. Even small changes in the radius can cause large changes in the flow rate; for example, a 10% increase in the radius causes the flow rate to increase by 46%! Table 2.1 gives examples of the dependence of flow rate on radius. Anyone familiar with syringes and needles is probably aware that the smaller the needle, the harder it is to get medication out of the syringe.

There are many applications of Poiseuille's law to the human body, and we will discuss some of these later. But I want to describe an example

TABLE 2.1. *Dependence of flow rate on the radius of a tube according to Poiseuille's law*

% change of radius	Multiplicative factor for radius	Multiplicative factor for flow rate	% change of flow rate
+10	1.1	$(1.1)^4 = 1.46$	+46
+20	1.2	$(1.2)^4 = 2.07$	+107
−10	0.9	$(0.9)^4 = 0.66$	−34
−20	0.8	$(0.8)^4 = 0.41$	−59

from meteorology, where this law can be used in a qualitative sense to describe wind (even though it is not moving through a tube). The pressure difference and the length in Poiseuille's law can be combined, so that we can rewrite the law as

$$Q = \frac{\pi r^4}{8\eta} \frac{\Delta p}{L}. \tag{10}$$

The term $\Delta p/L$ is called the pressure gradient. In this form, the flow rate is proportional to the pressure gradient. Perhaps you have heard meteorologists use this term. On a very windy day, often a meteorologist will say that there is a large pressure gradient. When a high-atmospheric-pressure system in a certain geographical area is located quite close to a low-pressure system, the pressure gradient is relatively large, causing the flow rate of air to be large, resulting in high winds.

Applications to the Human Body

It is now time to look at applications of pressure and fluids in the human body. The human body is quite complex and not all of the physics principles that we have discussed are perfectly applicable. Still, in many cases we can get excellent agreement with the ideas and equations presented. We will look first at the circulatory system and blood pressure, including the heart and blood flow, and include an application of giving medication into the blood stream. Then we will look at other parts of the body where pressure is important, such as the lungs, the eye, the brain, the bladder, bones, and a couple of other examples.

The Circulatory System and Blood Pressure

Usually when we think of the circulatory system, several things come to mind: the heart, arteries, and veins, and the flow of blood throughout the body. There are two major parts to the circulatory system: the *systemic circuit* and the *pulmonary circuit*. The systemic circuit is the primary system for delivery of blood to most of the body, with the pulmonary circuit used to replenish blood with oxygen in the lungs. In both cases, blood flows from the heart through arteries, capillaries, and veins, and then returns to the heart.

The circulatory system can be considered a closed system of tubes and pipes for most of our discussion. This concept is not entirely accurate because there is continual exchange of fluids into and out of the body, including the circulatory system. Also, arteries and veins are elastic, so that changes in their shape can affect flow at times. However, the physics concepts discussed will be applicable over short periods of time, or when averaging over longer periods of time. This idea of a closed system means that the flow rate of blood throughout the body is constant. The heart provides the starting pressure, acting like a pump, to push the blood through the system.

Because blood is a viscous fluid, it requires a continuous drop in pressure along the path of the circulatory system for blood to flow. In other words, there is a continual pressure difference exerted along the path of blood flow. In the systemic circuit, blood leaves the heart, passes through the aorta, then to the major arteries, through smaller arteries called arterioles, and then to the capillaries. On the return trip back to the heart, the blood flows from the capillaries, through the smaller veins called venules, then through larger veins, and finally through the vena cava before entering the heart. The pressure in the aorta is the highest along this path, with the lowest pressure at the vena cava.

From Poiseuille's law, the flow rate of a fluid is inversely proportional to the length of the tube. The systemic circuit is a longer pathway for blood to flow through because it is a much larger system. The pulmonary circuit is shorter because blood needs only to flow from the heart to the lungs and back to the heart. Thus the longer systemic circuit offers more resistance to blood flow, and therefore requires a higher starting pressure to maintain a certain flow rate. This pressure is in the range of 80 to 120 mm-Hg. The pulmonary circuit has lower resistance, so the starting pressure to maintain the same flow rate is lower, in the range of 10 to 25 mm-Hg.

By placing your fingers on your wrist or on your neck, you can feel your pulse if your fingers are pressing near a major artery. You can feel it because there is a small change in pressure that corresponds to the beating of the heart and causes the walls of the artery to expand and contract slightly. A similar example is observed if you place your hand on a balloon while someone else alternately squeezes it and releases it: you feel the balloon as it bulges outward slightly and then relaxes. The beating pulse is an example of Pascal's principle. If you measure the time for a certain number of beats, you can determine your pulse rate. Your pulse rate increases during exercise and decreases when at rest.

When you go to the doctor, you will often have your blood pressure checked. The pressure is reported using two numbers with the larger value stated first, followed by a smaller value. An example is "120 over 80," which is written "120/80." The units of these numbers are millimeters of mercury (mm-Hg). The pressure values are gauge pressures because the device used (a sphygmomanometer) has a gauge attached to it that reads zero when open to the atmosphere. Also, these pressures are measured in a major artery of the systemic circuit in the arm at about the same vertical level as the heart. Typical pressures in the pulmonary circuit range from 10 to 25 mm-Hg, written as 25/10, which is only a fraction of the pressure in the systemic circuit.

The larger pressure value occurs when the heart contracts. Consider Boyle's law for a moment, even though in this case we are applying it to a liquid instead of a gas. (This is not really appropriate because Boyle's law was developed to apply to gases; still, it illustrates the principle.) When the heart contracts, its volume decreases. This decrease in volume causes the pressure to increase. The larger pressure is called the *systolic pressure*. The term "systolic" comes from the Greek root word systole, which means "a drawing together or a contraction."[1] Systolic pressure is the blood pressure at contraction.

The lower pressure value occurs when the heart relaxes. During this time, the volume of the heart increases slightly, which results in a drop in pressure. The smaller pressure is called the *diastolic pressure*. The term "diastolic" comes from the Greek root word diastole, which means "a drawing apart."[2] Diastolic pressure is the blood pressure at relaxation.

Even though I used Boyle's law to help explain the changing blood pressure, it doesn't really apply to liquids. Another way to think about this issue is to consider Pascal's principle. The contraction of the heart can be thought of as the application of an external pressure to the enclosed blood. This additional pressure is distributed throughout the blood, which forces valves to open and close, resulting in blood that flows or stops flowing. Upon relaxation, the external pressure is removed and the blood pressure drops dramatically.

As was stated previously, the heart pumps blood through both the systemic and the pulmonary circuits. There are four chambers in the heart: the left ventricle, the right ventricle, the left atrium, and the right atrium (the plural of atrium is atria). Blood flows out of the left ventricle through the aorta and the systemic circuit, and returns to the right atrium. It then flows into the right ventricle, on through the pulmonary artery to the lungs, and returns to the left atrium via the pulmonary vein. The blood leaves the left atrium to enter the left ventricle, completing the trip through the two systems, and then is ready to start the journey again. A sketch of the heart that shows blood flow through its various parts is shown in figure 2.4.

It is the pressure changes as the heart relaxes and contracts that control the flow of blood through the heart. Blood flows into and out of the various chambers through four one-way valves. Two of these valves separate each atrium from its corresponding ventricle; they are called *atrioventricular (AV) valves*. The right AV valve is the *tricuspid valve* and the left AV valve is the *mitral valve*. The flaps of these valves hang loosely open when the heart is relaxed, allowing blood to flow into the two ventricles from the two atria. When the ventricles contract, the increased pressure pushes on these flaps, causing the two AV valves to close.

The other two valves, called *semilunar (SL) valves*, are the *aortic valve* and the *pulmonary valve*. Both of these valves are closed when the heart is relaxed. Upon ventricular contraction, the pressure in the ventricles exceeds the pressure in the aorta and the pulmonary artery, causing the two valves to open and allowing blood to rush into the aorta and the pulmonary artery. Upon relaxation, the pressure drops rapidly in the

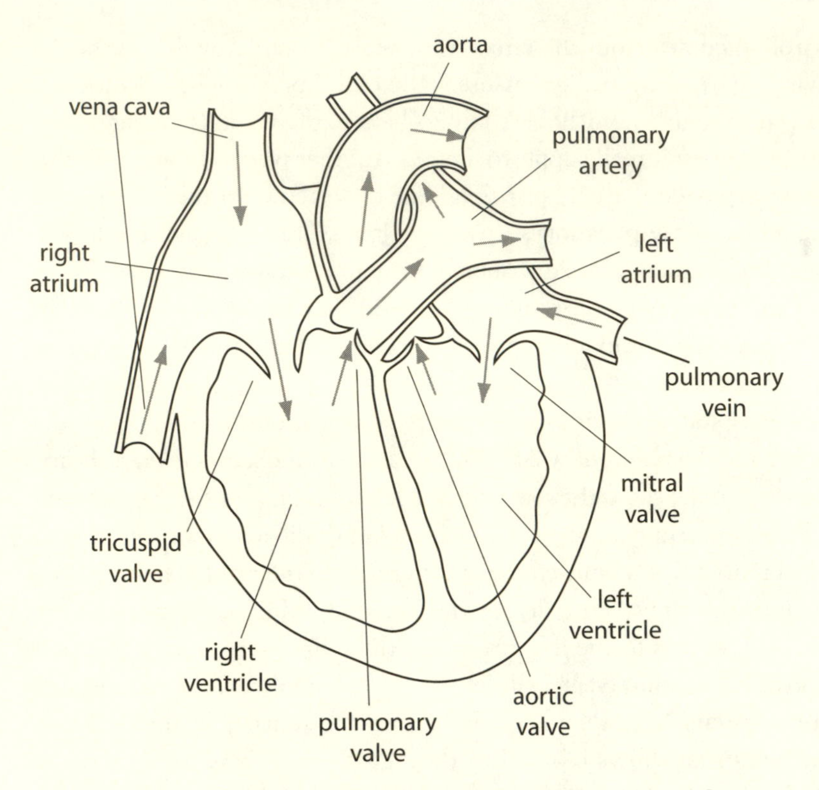

FIGURE 2.4. The heart and its blood flow. Oxygen-depleted blood flows into the right atrium and then through the tricuspid valve into the right ventricle. As the heart contracts, blood flows through the pulmonary valve into the pulmonary artery to the lungs. Reoxygenated blood from the lungs flows through the pulmonary vein into the left atrium, then through the mitral valve into the left ventricle. Again, as the heart contracts, blood flows through the aortic valve into the aorta to the rest of the body.

ventricles to nearly zero, allowing the blood in the two arteries to flow backward and causing the two valves to close.

Figure 2.5 shows a plot of the blood pressure in the aorta as a function of time over several cycles of the heart as it contracts and relaxes. The systolic pressure is the largest and the diastolic pressure the smallest value. The small rise in pressure after the systolic pressure has been reached occurs when the aortic valve closes. (This point in the graph is called the dicrotic notch.)

The familiar double-beat sound of the heartbeat corresponds to closing of valves in the heart. The first sound occurs at the beginning of the rapid rise toward the systolic pressure as the two AV valves close. The second sound occurs when the heart relaxes, resulting in the closure of the two SL valves.

The design of the heart is important to provide the proper pressure for these circuits. The left ventricle supplies pressure for the systemic circuit close to 120 mm-Hg. This value can be higher in patients with hypertension (high blood pressure). The right ventricle supplies the pressure for the pulmonary circuit, up to about 25 mm-Hg. Because of the higher pressure requirement in the systemic circuit, the left ventricle has a different shape from the right. The left ventricle is circular in cross section, which is important because a circular shape is ideal for spreading the higher pressure more evenly. In addition, the left ventricle has a thicker muscle wall that can withstand the higher pressure and can contract more so as to produce the higher pressure. The right ventricle is not circular, but more crescent shaped in cross section; the muscle wall is not as thick so there is less pressure generated in the pulmonary circuit.

The pressure must continuously decrease along the path of blood flow, so in the systemic circuit, the pressure drops off from the aorta, through the arteries, until it reaches the capillaries, where the pressure is about 20 to 30 mm-Hg. Blood continues to flow through the veins to the vena cava, where the pressure drops to near 0 mm-Hg. Figure 2.6 shows a plot of the blood pressure in the systemic circuit at various locations.

The reported values of blood pressure can vary quite a bit based on a person's position because pressure in a fluid can be caused by the weight

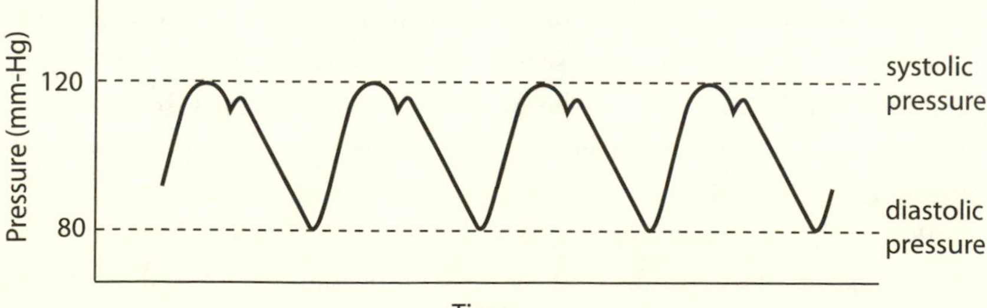

FIGURE 2.5. Blood pressure as a function of time over several cycles of the heart as it contracts and relaxes.

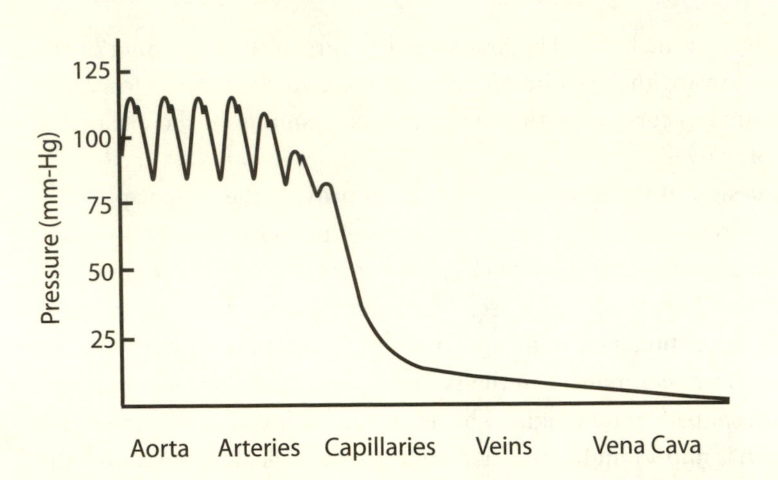

FIGURE 2.6. Typical blood pressure at various locations along the systemic circuit.

of the fluid [see eq. (1): $p = \rho g h$]. Thus, for a person standing, the pressure in the ankle can be much greater than at the aorta, and the pressure in the head can be smaller (fig. 2.7). For example, if the distance from the heart to the ankle is about 4 ft (1.2 m), then the increased pressure is nearly 100 mm-Hg:

$$p = \rho g h = (1050\ \text{kg/m}^3)(9.8\ \text{m/s}^2)(1.2\ \text{m}) = 12{,}300\ \text{Pa} = 93\ \text{mm-Hg}.$$

This increase can cause swelling in the ankles and feet if the person is standing for extended periods of time. If the individual elevates the feet, the additional pressure is relieved and the swelling diminishes.

Blood pressure is normally taken by placing a sphygmomanometer, or inflatable cuff, on the upper arm, at the same height as the heart. Thus, effects of vertical distance are minimized. But how is the blood pressure measured? The cuff is inflated so that the pressure is high enough to cut off blood flow in the arm (in the brachial artery). This pressure is larger than the systolic pressure. The nurse or doctor slowly releases the pressure in the cuff while listening for blood flow using a stethoscope placed on the inside bend of the elbow. When the pressure drops below the systolic pressure, blood begins to flow again, although the area of the artery is quite small. From the flow-rate equation (8), the speed of the blood is quite high, causing the flow to be turbulent, which produces a distinguishable sound. As the pressure applied by the cuff drops further, the area of the artery increases so that the flow slows enough to become laminar; the sound is no longer heard. The pressure where this occurs is the diastolic pressure. The cuff is a noninvasive method of measuring blood pressure. Invasive methods can be used, but they require inserting a needle, or cannula, directly into an artery.

Speed of Blood

Because the circulatory system is a closed system, the flow rate of blood should be constant throughout. This idea plays an important role in the body because, as the cross-sectional area of the circulatory system changes, the speed of blood in these different sections varies as well.

The cross-sectional area of the aorta in a typical adult is about 3 cm^2, and the speed of blood flowing through the aorta is about 30 cm/s, giving a flow rate of 90 cm^3/s. According to the flow-rate equation (8), or

$$A_1 v_1 = A_2 v_2$$

if the area were to increase, then the speed would decrease. This is exactly what happens along the systemic circuit. The total area of the arteries

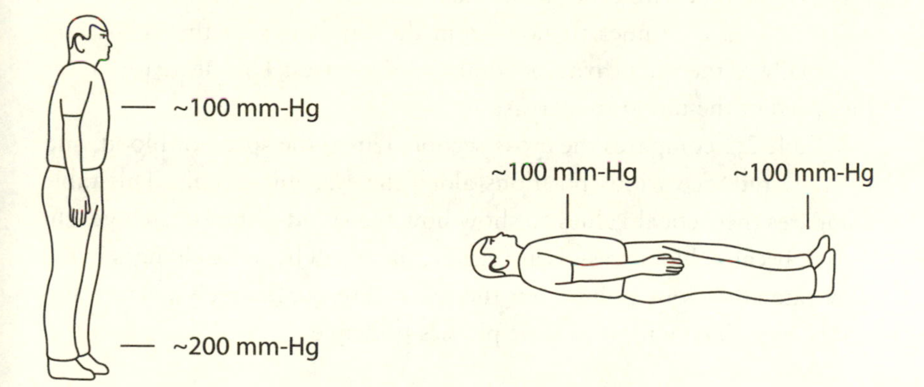

FIGURE 2.7. Dependence of blood pressure in the larger arteries on a person's position. If the person is standing, the pressure in the ankle includes pressure due to the weight of the blood; it can be quite high. If the person lies down, the additional pressure is relieved.

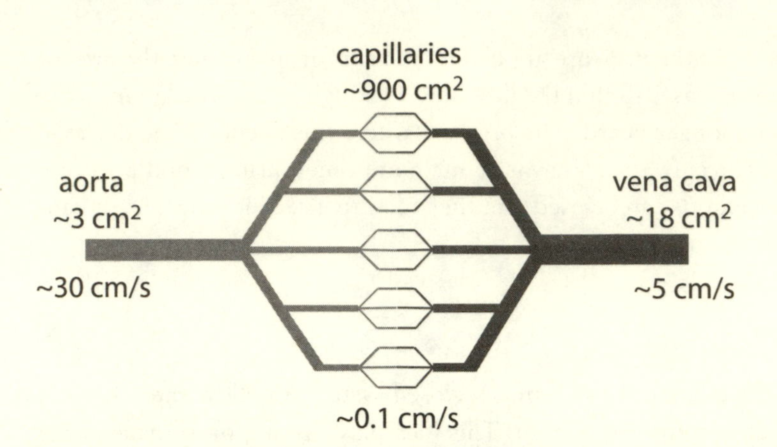

FIGURE 2.8. Dependence of blood velocity on area of blood vessels. The speed of blood in the aorta and the vena cava is higher than in the capillaries, because the area of the capillaries is much larger.

increases in comparison to the aorta, and the total area of the capillaries is larger again. Even though a single capillary is very small in diameter, there are so many of them (billions and billions) that the overall area is much larger (fig. 2.8). It is because of the large total area of the capillaries that the speed of blood flow in the capillaries is very small. This slow speed is important because that is where the exchange of gases occurs between the blood and the surrounding tissue.

As blood continues to flow from the capillaries to the veins, and eventually to the vena cava, the total area decreases. This decrease causes the speed of the blood to increase.

Table 2.2 compares the cross-sectional area, the speed of blood, and the flow rate at various positions along the systemic circuit. This table compares theoretical values to show how the speed of blood changes. In reality, because blood vessels are elastic, there will be some changes in the flow rate at various points, but the results are qualitatively correct and can be explained with this basic physics principle.

Example of Intravenous Medication

The following example integrates several ideas we have covered in this chapter. The problem is to find how high a bag of medication must be hung for it to be administered to a patient in a certain amount of time

TABLE 2.2. *Comparison of blood flow along the systemic circulatory system*

Location	Cross-sectional area	Speed of blood	Flow rate
aorta	3 cm^2	30 cm/s	90 cm^3/s
arteries	100 cm^2	0.9 cm/s	90 cm^3/s
capillaries	900 cm^2	0.1 cm/s	90 cm^3/s
veins	200 cm^2	0.45 cm/s	90 cm^3/s
vena cava	18 cm^2	5 cm/s	90 cm^3/s

though an intravenous (IV) tube. Figure 2.9 helps to illustrate this example. The medication enters the bloodstream through a needle inserted into a vein (usually in the inside bend of the arm, the top of the wrist, or the top of the hand). A vein is used because of the relatively low pressure, around 10 to 20 mm-Hg. If an artery were punctured, the pressure would be so large that blood would literally squirt out; external pressure would have to be applied for an extended period of time to get the bleeding to stop.

The medication is in liquid form, usually in a plastic bag, and we assume it has a volume of 500 cm^3 (500 cc). The bag is hung from a stand or pole placed a distance above the needle; a plastic tube runs from the bag to the needle.

Suppose that a 20-gauge needle (inside diameter of about 0.6 mm) with a length of 2 in. is used and we want it to empty in 30 minutes. How high must the bag be hung? We begin by calculating the flow rate

$$Q = \frac{\Delta V}{\Delta t} = \frac{500 \text{ cm}^3}{30 \text{ min}} = 0.278 \text{ cm}^3/\text{s}.$$

Then we use Poiseuille's law to calculate the pressure difference across the length of the needle. In this problem, the needle is the rate-limiting factor. In other words, the needle is so small compared to the size of the tube from the bag that the tube does not limit the flow, only the needle. When we solve for Δp in Poiseuille's law, we get

$$\Delta p = \frac{8\eta L Q}{\pi r^4}.$$

Plugging in the length and radius of the needle, along with the calculated value of Q, and using a value for the viscosity of the medication similar to that of water (0.001 Pl), we find that Δp = 4400 Pa = 33mm-Hg.

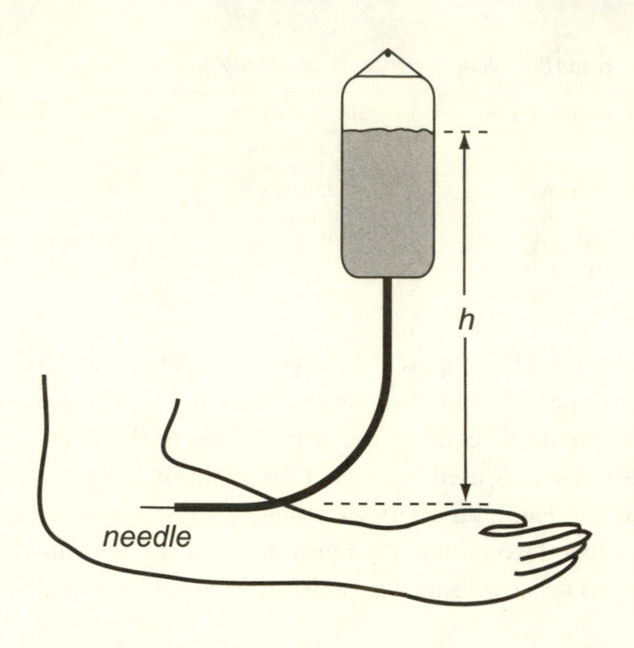

FIGURE 2.9. IV fluids given to a patient through a needle in a vein. The needle is at a depth of h below the surface of the fluid, so the pressure at the needle arises because of the weight of the fluid.

If the pressure in the vein is 20 mm-Hg, then the pressure at the inlet to the needle must be 33 mm-Hg higher, or 53 mm-Hg. This pressure is attained by placing the bag at a certain height above the needle given by

$$h = \frac{p}{\rho g}.$$

For an IV liquid that has a density very close to that of water (1 g/cm^3 = 1000 kg/m^3), the height is calculated to be about 0.7 to 0.8 m above the arm, which is consistent with typical heights used. Often, the bag is placed at an even higher position and a flow meter is used to regulate the rate at which the drug is administered.

There are several variations on this problem that could be considered. One is to give a blood transfusion instead of medication. Blood has a slightly higher density than water and is more viscous. Different needle sizes or flow rates could be considered as well. Another variation is to consider the case where the medication or a transfusion must take place quickly. In this situation, the flow rate should be much higher, which requires the pressure difference across the needle to be much larger. A practical way to increase the pressure is to squeeze the bag, applying Pascal's principle.

Example of Coronary Arteries

Poiseuille's law is also important in blockage of the coronary arteries. Coronary arteries deliver blood to the heart muscle so that it can function properly. If the coronary arteries are blocked, then blood cannot flow properly to the heart, the heart muscle does not receive proper nutrients and gas exchange, and a heart attack may occur. Narrowing of the arteries is caused by deposits of cholesterol on the inside walls of the arteries. Over time these deposits can build up to produce a blockage, as illustrated in figure 2.10.

Recall that, by Poiseuille's law, the radius of the tube is the parameter that the flow rate most strongly depends on. If the coronary arteries were to narrow by even a small amount, say a 10% decrease in the radius, then the flow rate would decrease by over 30% (see table 2.1). To compensate for this decrease in flow rate, the pressure difference across the length of the arteries may need to increase. This increase in pressure is one of the causes of high blood pressure.

There are several medical procedures that can be performed to restore blood flow to the heart muscle. One procedure is bypass surgery, which is performed by taking a larger vein from another part of the body, usually the leg, and attaching it to the blocked coronary artery so that blood bypasses the blocked portion of the artery. This procedure has been around for decades and is still in use now.

Another procedure that has become quite common is balloon angioplasty. In this procedure a long tube (catheter) is inserted into the coronary arteries and a balloon is inflated that compresses the cholesterol against the walls of the artery. Often a metal stent is left in place to help

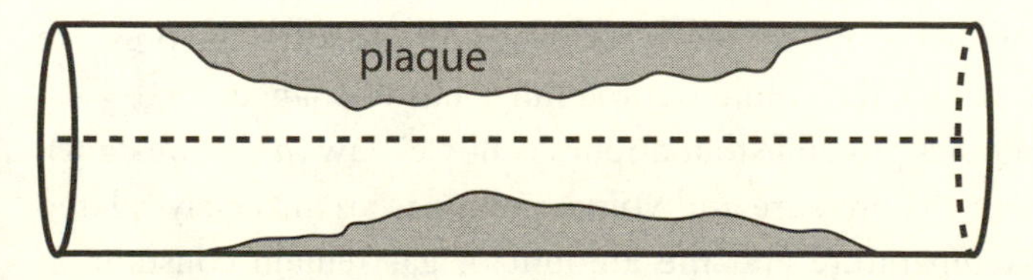

FIGURE 2.10. Narrowing of the arteries. Buildup of cholesterol, or plaque, on the walls of a coronary artery makes it harder to get blood through the artery.

keep the artery open. The stent is a gridlike array of metal, similar to chicken wire, which changes shape when the balloon is inflated.

Medication can be used to help dissolve the cholesterol on the walls of the artery, or to lower cholesterol levels in the blood to keep the arteries from becoming blocked. Some medications, called blood thinners, might be thought to lower the viscosity of blood, which from Poiseuille's law would be expected to increase flow rate. However, blood thinners *do not* change the viscosity and are simply used to keep blood from clotting; thus they do not change the flow properties of blood.

There are many more topics that could be explored related to blood pressure and blood flow, but now we move on to some other examples of pressure in the human body.

Pressure in Other Areas of the Body

The lungs, the eye, the brain, the bladder, and bones are other areas of the body where pressure plays an important role. To function properly, there has to be proper pressure in these areas of the body. Abnormal pressure can cause problems that may need a medical remedy.

The Lungs

During breathing, air goes into the lungs (*inspiration* or *inhalation*) and then leaves the lungs (*expiration* or *exhalation*). Because air is a fluid, there must be a pressure difference in order to move air into and out of the lungs. In inhalation, the pressure in the lungs must drop below atmospheric pressure, so that air from the outside will move into the lungs. In exhalation, the pressure in the lungs must be greater than atmospheric pressure, so that air will move out of the lungs. In order for the pressure in the lungs to change, the volume of the lungs has to change.

The basic physics principle that applies is Boyle's law (pV = constant). This law says that the pressure and volume of a gas are inversely related as long as the temperature and the amount of gas remain constant. If the volume of an enclosed gas were to increase, then its pressure would decrease; and if the volume were to decrease, then its pressure would increase.

In order for the pressure in the lungs to increase or decrease, the volume of the lungs, or more accurately the thoracic cavity, must change. For the volume to change, muscles in the thoracic cavity must move structures that are located around the lungs. These muscles include the diaphragm, which is a large, dome-shaped muscle attached to the bottom of the thoracic cavity. Other muscles can change the position of the rib cage and sternum so that the volume changes through their action.

In normal inhalation, the diaphragm moves downward, flattening out so that the volume of the thoracic cavity increases. Other muscles contract as well, which cause the rib cage to expand. The increase in volume causes the pressure in the lungs to decrease by about 4 mm-Hg below atmospheric pressure. If the trachea is open, air will flow into the lungs until atmospheric pressure is reached. At this point the inhalation process is complete. Deep breathing or forced inhalation, after exercise or for people with obstructive pulmonary disease, is produced by other muscles from the neck, chest, and back causing a larger volume, lower pressure, and faster air flow into the lungs. For someone with asthma, drugs like albuterol cause dilation of the airway to help with forced inhalation.

In exhalation, the diaphragm relaxes, moving upward toward the thoracic cavity, and other muscles relax so that the rib cage returns to its normal position. This action creates a smaller volume, thus increasing the pressure. The increased pressure (about 4 mm-Hg above atmospheric pressure) will cause air to move out of the lungs until atmospheric pressure is reached. Forced exhalation uses abdominal and other muscles to decrease the volume more rapidly, increasing the pressure, so that air can be expelled quickly.

In coughing and sneezing, the throat and/or nasal passages are closed so air cannot escape as the pressure increases dramatically. Once the muscles relax to open the passages, air rushes out quickly, often taking sputum or nasal discharge with it.

The Eye

Pressure within the eye is called *intraocular* pressure. Normal pressure values range from about 10 to 20 mm-Hg. If the pressure is too low, then the eye can collapse. If the pressure is too high, it can cause damage to the optic nerve; this is called glaucoma.

The eye has several important components: the *cornea*, the *aqueous humor*, the *lens*, the *vitreous humor*, and the *retina*. The importance of these components as they relate to optics of the eye is discussed in chapter 6. For discussion of pressure, these components play an important role as well.

The cornea is the outer surface of the front of the eye through which light travels to enter the eye. The aqueous humor is a clear fluid that fills the space between the cornea and the lens, with its supporting ligaments. The lens separates the front part of the eye from the rear part, which contains the vitreous humor. The vitreous humor is a clear gel-like liquid that helps maintain the shape of the eye and pushes on the retina to hold it against the back of the eyeball.

The aqueous humor is continuously produced, and continuously enters and drains from the front of the eye. In a normal eyeball, the production and drainage of the aqueous humor occur at the same rate. The vitreous humor is produced during the embryonic stage and is not replenished, but stays for the entire life of the person.

To measure pressure, a device called a tonometer is placed against the cornea and the pressure is based on how much force it takes to cause the cornea to flex, bend, or compress. Because the cornea is in contact with the aqueous humor, which is coupled to the vitreous humor, the measurement gives a value for the intraocular pressure. Pressure values have to be interpreted carefully because factors such as rigidity and thickness of the cornea can affect them.

As stated previously, production of the aqueous humor normally occurs at the same rate as drainage. However, if drainage is limited or blocked for some reason, additional aqueous humor in the front of the eye can cause the pressure to build up. The higher pressure can cause the vitreous humor to push harder on the retina, pushing on the blood vessels that feed the retina and the optic nerve. This condition can cause damage to the eye and may lead to blindness. Damage to the optic nerve is called *glaucoma*, with the major contributing factor being increased pressure in the eye.

The usual treatment for high intraocular pressure is medication, in the form of eye drops, that reduces the rate of production of aqueous humor (such as timilol, a beta-antagonist) or that increases the outflow of aqueous humor (such as cholinergic drugs or alpha-antagonists). The lower pressure helps to reduce the possibility of damage to the optic nerve. Other treatments include laser surgery and conventional surgery designed to increase the drainage of aqueous humor.

The Brain

The brain is part of the central nervous system, which also includes the spinal cord. The brain functions as the electronic control system and memory storage device for the body. The complexity and flexibility of the brain are remarkable. Because of its importance to the body, the brain is protected by several layers, which include the skull, a set of tissue membranes called the meninges, and cerebral spinal fluid.

The *skull* is a complex structure that is composed of the cranial bones, or cranium, and the facial bones. The cranium not only protects the brain, but also contains the brain. Thus, if any fluid were to build up or if any swelling were to occur, the pressure in the brain would increase.

The *meninges* are a set of three membranes that cover the brain and contain cerebral spinal fluid. They connect the brain to the skull, keep various parts of the brain separated, and control the flow of cerebral spinal fluid.

The *cerebral spinal fluid* (CSF) is a watery-like fluid that surrounds the brain and the spinal cord. It provides a layer of cushioning, allowing the brain to float within the skull, and thus provides mechanical protection to the brain. The composition of CSF also provides protection through regulation of blood flow and distribution of hormones. The brain contains about 150 mL of CSF at any time. It is continuously produced, and must be continuously drained as well.

Pressure in the brain, *intracranial* pressure, arises from several sources: the pressure that the skull exerts on the brain and the presence of fluids such as blood and cerebral spinal fluid. Normal pressure in the brain is around 50 to 180 mm-H_2O (which corresponds to about 4 to 13 mm-Hg), but can vary based on position and activity of the person.[3,4]

Elevated pressure in the brain can cause damage to the brain. Traumatic injuries that cause swelling, tumors that increase volume or that block flow of CSF, diseases that affect flow of CSF, and high blood pressure can cause increased intracranial pressure. In order to decrease the pressure, medical treatment must be performed.

One of the most common causes is overproduction of CSF or blockage of the drainage system (similar to the situation that can occur in aqueous humor in the eye). The abnormal accumulation of CSF in the brain is called *hydrocephalus*. In newborns, the increased pressure can cause the head to be enlarged because the cranial bones have not completely fused. This makes the cartilage between the cranial bones still malleable. In older children and adults, this condition can lead to brain damage and many neurological problems, including headache, double vision, and changes in personality. The usual procedure to treat hydrocephalus is to surgically implant a shunt system. This system drains trapped CSF to another part of the body where it can be absorbed.

Traumatic head injury can rupture blood vessels (hemorrhage) or cause swelling, both of which can increase the intracranial pressure. To treat a hemorrhage, surgery is required to remove the quantity of blood and to repair the blood vessels. Swelling of the brain is treated by medication to reduce inflammation in the damaged tissue or by decreasing the blood volume to the brain.

The Bladder

The urinary system is composed of several organs including the kidneys and the bladder. The kidneys filter waste from the blood and produce urine that carries these wastes from the body. The bladder serves as a temporary storage device to hold the urine until it is excreted through the urethra. Sphincter muscles control when urine is able to pass from the body.

The bladder is a collapsible, hollow organ in which urine collects. When the bladder is empty it occupies a relatively small volume, but when it fills it expands to as much as 500 mL or more in volume. Pressure in the bladder increases when it begins to fill. When about 200 mL of urine has collected, the bladder has stretched to a point where certain nerves transmit a signal to the brain of the need to urinate. If the person does not empty the bladder at that time, usually the urge subsides for a time until another 200 mL or so of urine collects and the bladder is full.

Pressure in the bladder when it is empty or nearly so ranges from 0 to 10 cm-H_2O or about 7 mm-Hg (gauge pressure). As the bladder fills, the micturition or urination ("gotta go") pressure is above about 30 cm-H_2O (22 mm-Hg).[5]

Two examples of the effects of pressure on the bladder are related to women and pregnancy and to men and an enlarged prostate gland. Pregnant women often indicate they have to go to the bathroom quite often. A simple application of Pascal's principle helps us understand why. When the unborn child becomes sizable inside the mother, the baby will push (unknowingly of course) on the mother's bladder. This external pressure applied to the enclosed fluid (Pascal's principle) causes the pressure to increase throughout the bladder, stretching it enough that the micturition pressure is exceeded. Thus, the brain receives the signal that it is time to go. Sometimes the pressure is large enough that the urine goes past the sphincter muscles involuntarily. The inability to control urination is called *incontinence* and is not normal for adults.

In men, the prostate gland surrounds the urethra coming from the bladder. If the prostate gland is enlarged, it may press on the urethra causing the tube to narrow so much that the bladder is not able to empty completely. In some cases, the pressure in the bladder can be as high as 100 cm-H_2O (about 70 mm-Hg). In order for urine to flow, it must overcome the pressure caused by the prostate. The bladder is usually not able to empty before flow is stopped. The bladder then fills again, and the man feels the urge to urinate again. The inability of the bladder to empty completely is called *urinary retention*. It is usually treated by special drugs that dilate the urethra, increasing its radius and so decreasing the pressure required to keep the urethra open.

Bones

Pressure in solids is often called stress. You are probably most familiar with this term as it relates to being overworked, tired, or too busy. You may have heard the term "stress fracture," which can occur in a bone through application of repetitive forces. It occurs most often in the foot or ankle and is seen in athletes who run or jump on hard surfaces. More severe fractures or breaks occur when much larger forces are applied over a very short time, as in a car accident or a fall.

Bones and joints are subject to large pressure. For example, suppose

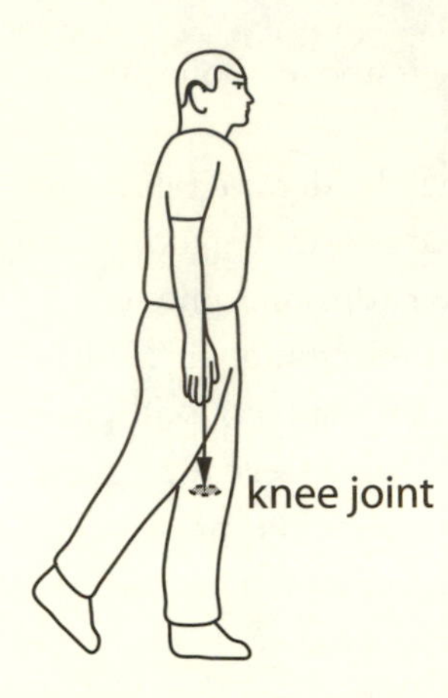

FIGURE 2.11. Pressure on the knee joint caused by the weight of a person acting on the small area of the joint.

that a person weighing 180 lb stands on one leg, as shown in figure 2.11. If the knee joint has an area of 0.5 in.^{2}, then the pressure is 340 lb/in.^{2} or about 23 atm (only about 94% of the weight acts on the knee joint; 6% is below the knee). For someone who is running, there are larger pressures applied to the joint.

Normal bones break when the applied pressure is greater than about 15,000 lb/in.^{2} or 100 atm. When a football player tackling a running back applies his weight (or more) to the shin bone, the force may be enough to break the bone. In a car accident, if the driver's skull strikes the steering wheel, the skull may be fractured.

Bone mineral density, or bone mass density, is a measure of bone strength and the ability of bones to bear weight or withstand pressure. There are two main types of bone density measurement, dual energy x-ray absorptiometry (DEXA) and quantitative computed tomography (QCT). DEXA measures areal density, the density in a cross-sectional area of the bone. QCT measures volume density. Measurements are usually made at the spine or the hip, although sometimes they are made at the heel, finger, or wrist.

As a person ages, the amount of calcium in the bones decreases and the bones become more fragile. Peak bone density occurs at about 20 years of age. *Osteoporosis* occurs when the bone density is significantly reduced and the individual is at a high risk of bone fracture. *Osteopenia* is an early stage of osteoporosis. The ability of materials to withstand fracture is related to stress and strain through Young's modulus or the bulk modulus.

Other Examples

There are many other examples of pressure in the human body, but I will mention briefly just two more. One has to do with scuba diving and the other involves cutting tissue.

A scuba diver experiences large pressure (due to the weight of the water) when he or she dives below the surface. This pressure acts on the entire body, including the lungs and blood. All pressures on a person depend not only on atmospheric pressure when out of the water, but also on the ρgh pressure due to the weight of the water. When a diver rises to the surface of the water, the air in the lungs expands, which requires the diver to continually exhale when surfacing. Also, small bubbles of nitrogen and other inert gases in the bloodstream expand when the diver rises, so that they can block the flow of blood in the circulatory system, causing intense pain, severe illness, or death. This condition is called *decompression sickness* (the bends), and, in order for divers to avoid this problem, they must pause for several minutes at various points during their ascent for proper exchange of gases to occur.

For a knife to cut tissue (by accident or in surgery), the force exerted on the tissue does not have to be very large. If the knife is very sharp, then the surface area of the blade is very small. A reasonable amount of force applied over the very small area can result in a huge pressure, which will be enough to cut the tissue. As an example, suppose you were to apply only 1 lb of force to the tip of a knife that is only 0.5 mm in diameter. The resulting pressure is about 3200 lb/in.^{2} or about 210 atm, which is much more than adequate for cutting tissue.

Summary

Pressure is an important concept in the body. Blood pressure is required to cause blood to flow so that it can carry out its many functions in the body. Other organs and systems of the body use the presence and/or movement of fluids to serve particular functions. Even solid objects in the body, such as bones, are subjected to pressure. We have looked at some basic concepts about pressure in the body, exploring a few applications in a bit more detail. Many of these areas are studied in depth by healthcare professionals and scientists so that when health problems arise in individuals they can be treated properly.

THREE

Energy, Work, and Metabolism

Energy is important in many aspects of everyday life. Our cars need energy so that we can get around easily. Our homes use energy to stay warm in winter and cool in summer. Our bodies need energy to move, to work, and to stay warm. Sometimes we are too warm and we need to remove energy so we can cool off. We get the energy for our bodies from the food we eat. If we eat too much, then our bodies can store excess energy in fat.

For all the processes connected with our bodies, energy is required. At the macroscopic level, we need energy to run, throw a ball, and climb stairs. At the microscopic level, energy is required for chemical reactions to occur, for moving cells or parts of cells, and even for feeling pain. *Energy is defined as the capacity to do work*. In this chapter we will see how work and energy are related to the human body and look at topics such as food calories, regulation of body temperature, dieting, and exercise, to name a few.

Work

As I just said, energy is defined in terms of its capacity to do work, but, as so often, this definition is not the complete story. Actually, objects possess energy, which can take several different forms. This energy can be used by the object to perform work on another object. Work occurs when a force is applied by one object to another object and the second object moves. Work is done only if there is motion involved when a force is acting. When you throw a baseball or kick a football, you are doing work on the ball.

Work requires the presence of a force. You may recall from Newton's first law of motion that an object can be moving even if there are no forces acting on it; no work is done in this situation because work requires a force. However, if the object was initially at rest, then to get it moving, a force had to be applied to it; work was done on the object to get it to start moving. To get the object to stop moving, work has to be done on it as well.

Suppose your car is out of gas and you have to push it to the nearest gas station. If you push on the car and the car begins to move, then you are doing work on the car. If you push on the car and it doesn't move, then you are not doing work on the car. You may get quite tired, but there is no work done on the car. There is, however, work going on in your body, which is why you get tired.

Work done on an object by a force can be positive, negative, or zero. When you throw a baseball, you do positive work on the baseball, but when you catch a baseball, negative work is done on it. The positive or negative value of the work depends on the direction of motion of the object relative to the direction of the force acting on it. If the motion is in the same direction as the force, or even partly so, then positive work is done. If the motion is in the opposite direction, or partly so, then negative work is done. If the motion is in one direction, but the force is perpendicular to the motion, then no work is done. This occurs when you walk on a horizontal surface while holding a baseball in your hand—you apply an upward force on the ball, but the motion is sideways, so you do no work on the ball.

We define the work W quantitatively as the product of the size of the applied force F, the distance that the object moves, d, and the cosine of the angle θ between the direction of the force and the direction of the motion. It is written

$$W = Fd \cos \theta. \quad (1)$$

The angle θ ranges from 0° to 180°. If the angle between the applied force and the distance moved is between 0° and 90°, the work is positive (because the cosine is positive for angles in that range). If the angle is between 90° and 180°, the work is negative (because the cosine is negative

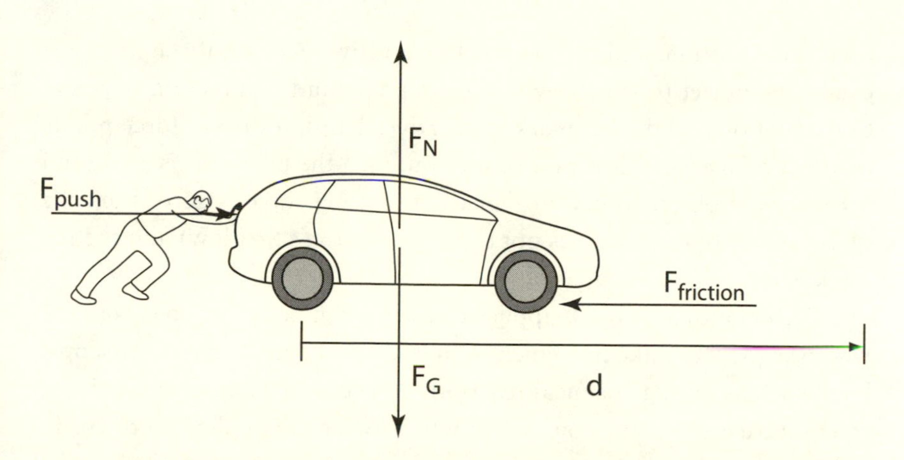

FIG. 3.1. Work done on a car as it moves to the right through a distance d. The push force F_{push} does positive work and friction $F_{friction}$ does negative work. The normal force F_N and the weight F_G do not do any work because they make a 90° angle with the direction of motion.

for angles in that range). For the case where the angle is exactly 90°, the work is zero because the cosine of 90° is zero.

Think again about the stationary car. You may recall from chapter 1 that there are several forces acting on the car: your push force in the direction of movement of the car, friction acting in the opposite direction, and two other forces: the downward weight and the upward normal force (fig. 3.1). So, based on our definition of work above, you do positive work on the car, friction (supplied by the ground) does negative work, and the weight and normal forces do no work because their angles are at 90° to the car's direction of motion.

When several forces act on an object, it may be possible to calculate the work done by each force individually. If these are added together, we have what is called *net work*. Mathematically, we write this definition as

$$W_{net} = W_1 + W_2 + W_3 + \cdots . \qquad (2)$$

The net work can also be determined from the definition of work above, but using the net force in the formula, or

$$W_{net} = F_{net} d \cos \theta. \qquad (3)$$

There are several implications of this last expression. First, if the net force causes the object to speed up, then it points in the general direction of the motion, and the net work is positive. Second, if the net force causes the object to slow down, then it points in the direction opposite to the motion and the net work is negative. Third, if the net force points perpendicular to the direction of motion, then the net work is zero (and it causes the object to curve). Finally, if the net force is zero, then the object's state of motion does not change (Newton's first law) and the net work is zero.

The definition of work applies to the human body for objects at the macroscopic level, like movement of a body part, and at the microscopic level, such as during chemical reactions and cellular motion. In chemical reactions, electrons and ions move when forces are applied to them. In cellular motion, components move into and out of the cell and individual cells move when forces act on them.

Energy

There are two types of energy related to the motion and interaction of objects: *kinetic* and *potential* energy. There are other types of energy, many of which are based on these two forms. Let's look at kinetic energy and potential energy in more detail, and then we will define some other types of energy.

Kinetic Energy

Kinetic energy is the energy of motion. If an object is moving, it possesses kinetic energy. Kinetic energy depends on the mass of the object and the speed of the object. It is expressed mathematically as

$$K = \frac{1}{2}mv^2, \qquad (4)$$

where K is the kinetic energy of the object, m is its mass, and v is its speed. The more massive an object, the more kinetic energy it has. A bowling ball has more kinetic energy than a beach ball moving at the same speed. Also, the faster an object moves the more kinetic energy it has—a 100 mph fastball has more kinetic energy than a 60 mph curveball (fig. 3.2).

Kinetic energy has the capacity to do work if the moving object pushes

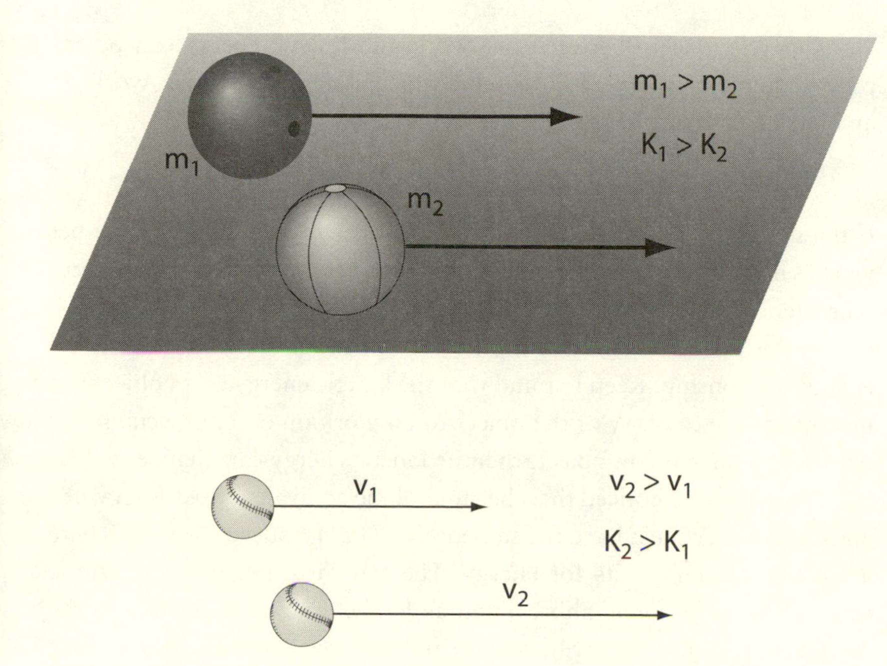

FIGURE 3.2. Kinetic energy. If a bowling ball and a beach ball move at the same speed, the bowling ball has more kinetic energy K_1 because its mass m_1 is larger than the mass m_2 of the beach ball, which has kinetic energy K_2. For two identical baseballs, the one that is moving faster at v_2 has more kinetic energy K_2 than the other one moving at v_1, which has kinetic energy K_1.

or pulls on another object. A moving hammer can drive a nail, a wrecking ball can knock a building over, and a moving car can bend a guard rail. Each of these objects has kinetic energy that does work—not all work is useful work, but it is still work.

Not only can objects with kinetic energy do work, but when net work is done on an object, its kinetic energy changes. If net work is done on an object and that net work is positive, the object will speed up; if it speeds up, its kinetic energy increases. This is what happens when you throw a baseball. If the net work done on an object is negative, the object slows down, as when you catch a baseball. If no net work is done on an object, it does not speed up or slow down, so its kinetic energy does not change. If you are pushing a car and it is moving at constant speed, then the work done by you is positive and cancels the negative work done by friction; the net work is zero and the kinetic energy of the car does not change.

This relationship between net work and kinetic energy can be expressed as the *work-energy theorem*, which states that the net work on an object equals its change in kinetic energy, or

$$W_{net} = \Delta K. \quad (5)$$

If the net work is positive, then the kinetic energy increases. If the net work is negative, then the kinetic energy decreases. If the net work is zero, then there is no change in the kinetic energy.

At this point, you may be a bit confused about the whole work and energy relationship. Keep in mind that the kinetic energy of an object is a measure of the capacity of the object to do work on other objects. But if net work is done on an object, then its kinetic energy can change.

You may have noticed that, because of the equivalency between work and energy, they must have the same units. That is indeed the case. There are many different units for energy. The standard unit in the scientific community is the *joule*, abbreviated as J. The joule is equivalent to the newton (force) times meter (distance) and to kilogram (mass) times meter squared per second squared (speed squared). In symbol form we write

$$1\ \text{J} = 1\ \text{N m} = 1\ \text{kg m}^2/\text{s}^2.$$

Other units include the erg and the foot-pound. We will come across even more units when we talk about heat exchange between objects and energy storage in food, such as the calorie, the kilocalorie, the Calorie, and the British thermal unit, or BTU.

Potential Energy

Potential energy is the energy of position or the energy of configuration. The potential energy of an object arises as it interacts with other objects around it through gravity, chemical bonds, or electrical forces. An object with potential energy has the *potential* to do work even if it is not currently doing work. Another way to think of potential energy is that it is stored energy or stored work.

Types of potential energy include gravitational, elastic, and electric potential energy. Gravitational potential energy of an object depends on its vertical position relative to the surface of the earth; it is caused by the force of gravity acting on the object. Elastic potential energy occurs in a

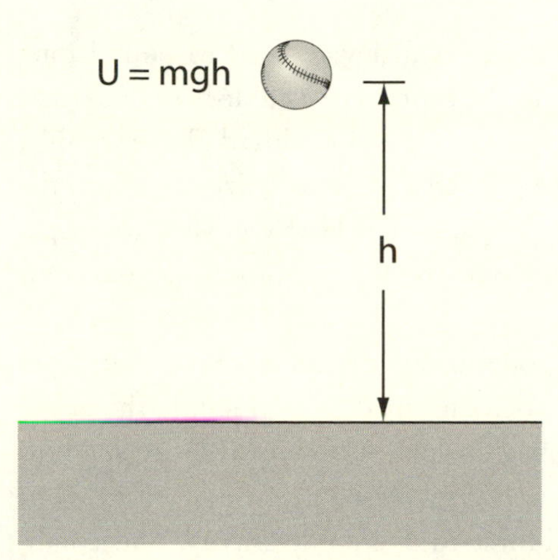

FIGURE 3.3. Potential energy. The gravitational potential energy U of an object, such as the baseball shown, depends on its mass m, the acceleration due to gravity g, and the vertical position h from a reference point, often taken as the ground.

spring and depends on how much the spring is stretched or compressed from its resting or equilibrium length; it is caused by an elastic force that tries to restore the spring to its original length. Electric potential energy of a charge depends on its position with respect to other charges; it is caused by electric forces acting on the charge exerted by other charges.

Each type of potential energy can be quantified, but I want to discuss the formula for gravitational potential energy because it is most closely related to (macroscopic) motion of the human body, particularly when moving around on the earth. The gravitational potential energy U depends on the mass of the object m, the acceleration of gravity g, and the vertical position h relative to the earth, and is written

$$U = mgh. \quad (6)$$

The parameter h can be interpreted in several different ways, but for our purposes we will assume it is the vertical height of the object measured from a reference point, often taken as the surface of the earth (fig. 3.3). The point of the formula is to show that the higher the object the larger the gravitational potential energy. An object 10 m above the ground has ten times more potential energy than an object only 1 m above the ground. The object located 10 m above the ground has the potential to do more work than the object located only 1 m above the ground.

So how is potential energy related to work? If an object is dropped from a certain height, the potential energy is converted into kinetic energy, and if it comes into contact with another object it can do work on that object. For example, a pile driver is a device used to pound long wooden, concrete, or steel poles into the ground to establish a foundation for a large building or in poor soil. A large weight is lifted above the pile, released, and then falls toward the pile, striking it and driving it into the ground. Another, more painful, example would be if you were to drop a bowling ball on your foot; the ball would exert a large force on your foot, perhaps causing it to be crushed.

Not only can the object with potential energy do work, but work can be done on the object. Actually, when an object is in free-fall, the gravitational force does work on the object to pull it downward, causing it to speed up, thus changing its kinetic energy. Also, to lift an object, work must be done on it by the applied force to overcome gravity. This condition is important for the human body—in order to lift an object or even to climb stairs, the body must expend energy to overcome the force of gravity and do work on the object that moves upward, whether the object that is being lifted or the body itself. This concept is important when we get into the area of exercise and food calories.

Other Types of Energy

There are many other types of energy, several of which are based in some way on kinetic and potential energy.

Mechanical energy E is defined simply as the sum of the kinetic and potential energy, written as

$$E = K + U. \qquad (7)$$

It describes the energy of an object in motion and takes into account the energy associated with its position or configuration. Suppose you drop a ball from a certain height with the ball initially at rest so that gravity is the only force acting on it. The ball starts out with no kinetic energy, so the mechanical energy is equal to the initial gravitational potential energy because the K term in equation (7) is equal to zero. As it falls, its potential energy decreases while its kinetic energy increases. The instant just before strikes the ground, its potential energy is zero and the mechanical energy is equal to the kinetic energy only because the U term in equation (7) is equal to zero. This situation is an example of a principle called the *conservation of mechanical energy* where the mechanical energy remains constant throughout the process even though the potential and kinetic energies change; the decrease in potential energy turns out to equal the increase in kinetic energy. When the ball comes into contact with the ground, the mechanical energy is no longer constant because there is now another force (the normal force) exerted by the ground on the ball, causing it to slow down and stop.

Electrical energy is related to the motion and position of electric charges, so it involves kinetic and potential energy as well. I have already mentioned the electric potential energy that results from interaction of charges with each other. If two positive (or two negative) charges are held near each other and then released, they fly apart because they repel each other. The forces they exert on each other cause them to move. The potential energy that they had initially gets converted to kinetic energy. However, a negative charge and a positive charge will move toward each other because they are attracted to each other—the negative charge pulls on the positive charge and the positive charge pulls on the negative charge (recall Newton's third law—the law of action-reaction). Thus, for like charges, the potential energy is highest when the distance between them is very small; for unlike charges, it is highest when they are very far apart.

Internal energy is associated with the motion and position of atoms and molecules that make up an object. If a baseball is at rest on the ground, it has no kinetic energy and no gravitational potential energy from a macroscopic perspective. However, because the atoms and molecules in the baseball are moving and interacting with each other, it has internal energy. This internal energy is a combination of kinetic and potential energy, but on a molecular and atomic scale.

Heat is energy that is in transit from one object to another object because of a temperature difference. Heat is transferred from the warmer object to the cooler object. When heat is added to an object, its internal energy increases so that its molecules and atoms have more energy. This may cause an increase in temperature of the object, or the object may undergo a phase change (such as when water boils). When heat is removed from an object, its internal energy decreases, causing it to cool or go through a phase change (such as when water freezes).

Chemical energy is potential energy that is stored in the bonds of chemicals. Because of the electrical forces acting between atoms, electrons, and protons, energy is required to maintain their configurations. When a chemical reaction occurs, some of this energy may be released as kinetic energy, causing the object to move or to heat up. A particularly rapid chemical reaction may cause an explosion, as when gasoline burns. Food has chemical energy that is released so that we can walk, run, lift objects, and perform work. Within our bodies, this energy is used to move cells and maintain various internal functions.

Other types of energy include *electromagnetic energy*, like the visible light needed for sight, ultraviolet radiation that causes sunburns, and x-rays used for medical diagnosis. Energy in sound waves (*acoustic energy*) is used for audible communication; these waves are produced when we speak by vibrating vocal cords and are transformed to mechanical vibrations in the ear that allow us to hear. *Nuclear energy*, in the form of radiation emitted by the nucleus of an atom, can be used to treat cancer (or can even cause cancer).

Conservation of Energy

When energy principles are discussed, it is important to understand the concept of *open* and *closed* systems. An open system is one where energy and matter can be transferred into or out of the system; a closed or isolated system does not allow for this.

One of the fundamental principles of physics is that energy is conserved, which means that the total energy in a closed system remains constant. Energy can be converted from one form to another, but the total amount of energy remains constant. If a system is closed or isolated it does not interact with anything outside of itself. If a system is not closed then energy can move into or out of it. Often, if a system is not closed, we can enlarge it to include those objects that exchange energy with the original system. Ultimately, the entire universe can be considered a closed system, so that the total energy of the entire universe is constant.

The principle of conservation of energy is important to the human body. Most of the types of energy mentioned above are used by the human body. Energy from food in the form of potential energy released from chemical bonds allows us to do mechanical work. Our bodies maintain a fairly constant temperature in order to function properly. If we exercise vigorously, we may produce excess heat and our bodies must get rid of it somehow. If we eat too much, the food energy is stored in fat tissue. All the energy that is used by our bodies can be accounted for as they perform their functions and interact with other objects around us.

Power

Power is defined as the rate at which work is done. Mathematically, it can be expressed as

$$P = \frac{W}{t}. \tag{8}$$

With work expressed in joules and time in seconds, the unit for power is the joule per second, which is renamed the watt, abbreviated as W. Thus, we see that 1 J/s = 1 W. There are other units as well, such as horsepower for motors and engines, where 1 hp = 746 W, and kilocalories per second or minute or hour when discussing food intake or metabolic rate for the body.

While power is defined in terms of work, it is also used when discussing the rate at which energy is converted from one form to another. For example, a 100 W light bulb converts electrical energy into light and heat at the rate of 100 J/s. The power of large electrical appliances like refrigerators is often listed in units of kilovolt-amps; 1 kilovolt-amp is equivalent to a kilowatt.

Efficiency

Another way to think about the conversion of energy is in terms of efficiency. Efficiency is usually defined in terms of work output divided by energy input, or

$$\text{Efficiency} = \frac{\text{Work output}}{\text{Energy input}} \times 100\%. \tag{9}$$

Motors, machines, appliances, and other devices are usually rated according to efficiency. It is a statement of how good a device is at accomplishing its task compared to how much energy is supplied to the device.

Efficiency can also be defined in terms of power: by comparison of the input power P_{in} to the output power P_{out}. Mathematically, this is written

$$\text{Efficiency} = \frac{P_{out}}{P_{in}} \times 100\%. \tag{10}$$

For the human body, efficiency is typically less than 20%, depending on the activity. A person doing work with an efficiency of 20% would require five times as much energy as the work being performed. The energy that does not go into useful work is typically converted into heat energy. By contrast, electric motors have efficiencies in the range of 80 to 95%, although automobiles have efficiencies around 20%, similar to the human body.

Temperature and Heat

Let's look more closely at heat energy and the concept of temperature because these concepts are important for proper function of the human body.

Temperature

Temperature is a measure of the relative hotness or coldness of an object. If one object feels hotter than another object, then the first object is at a higher temperature than the second. Because temperature is a relative measure, we use temperature scales to help us quantify the measurement. Common scales include the Fahrenheit scale, the Celsius (or centigrade) scale, and the Kelvin scale. Normal body temperature is 98.6°F, or 37°C, or 310 K. Water at atmospheric pressure freezes at 32°F, or 0°C, or 273 K, and it boils at 212°F, or 100°C, or 373 K(fig. 3.4). To convert between a Fahrenheit temperature T_F and a Celsius temperature T_C, we use the equations

$$T_F = \tfrac{9}{5}T_C + 32 \tag{11}$$

and

$$T_C = \tfrac{5}{9}(T_F - 32). \tag{12}$$

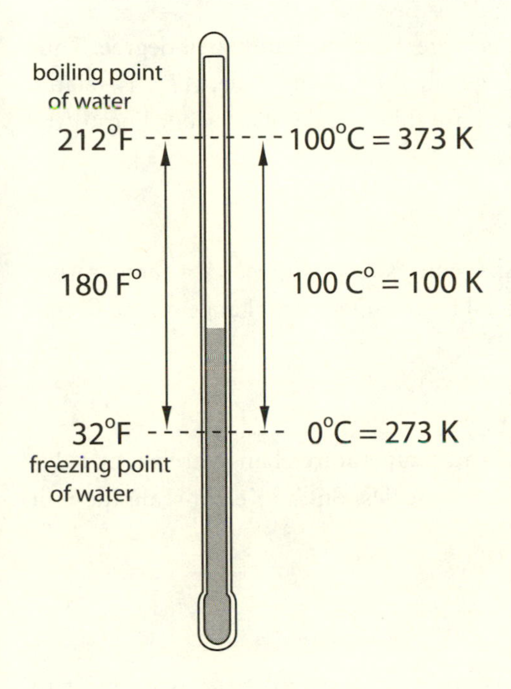

FIGURE 3.4. The thermometer indicates the temperature of an object. Shown here is a comparison of the boiling point and freezing point of water on the Fahrenheit (°F), Celsius (°C), and Kelvin (K) scales. Also shown is a comparison of the temperature *difference* between these two points measured in F°, C°, and K.

To convert between a Celsius temperature T_C and a Kelvin temperature T_K, we use the equation

$$T_K = T_C + 273. \tag{13}$$

The Kelvin scale is an absolute scale, which means that it starts at zero and has no negative values. Early evidence of this absolute scale came from studies of gases that measured the pressure of an enclosed gas as the temperature decreased. As the temperature drops, it is found that the pressure falls as well. At some point, the gas condenses to a liquid and the pressure drops to zero. However, above this point the pressure shows a linear dependence on temperature. If a straight line is drawn and extended to where the pressure drops to zero, this is found to occur at –273°C. This point became absolute zero on the Kelvin scale.

Temperature Change

The sizes of a Fahrenheit degree, a Celsius degree, and a Kelvin degree can be compared. Because there are 212 – 32 = 180 Fahrenheit degrees (F°) between the boiling and freezing points of water, but only 100 Celsius degrees (C°), the Celsius degree is larger than the Fahrenheit degree. Thus we see that a temperature change on the Celsius scale, ΔT_C, is related to a temperature change on the Fahrenheit scale, ΔT_F, according to the expression

$$\Delta T_C = \tfrac{5}{9}\Delta T_F. \qquad (14)$$

A Celsius degree is the same size as a Kelvin degree, so a temperature change on the Celsius scale is equal to a temperature change on the Kelvin scale, ΔT_F, or

$$\Delta T_C = \Delta T_K. \qquad (15)$$

Although units for temperature and temperature change are slightly different for the Fahrenheit and Celsius scales, on the Kelvin scale the unit is K for both.

Heat

We have already defined heat as energy transferred from one object to another because of a temperature difference. As heat leaves one object, its internal energy decreases, and as it enters another object, its internal energy increases. The heat flows between the objects until the temperatures of the two objects are the same. That doesn't mean that the internal energy of each is the same, because internal energy depends on the *type* and *amount* of material in each object.

If it is possible for heat to flow between two objects, we say that they are in *thermal contact*. They do not have to be in physical contact, as when heat flows from the sun to the earth. When the two objects are at the same temperature, there is no net heat transfer between them; they are in *thermal equilibrium*.

For example, suppose you have a cold can of soda sitting on a table in your room. The soda is most likely colder than your room, so heat will flow from the room into the soda causing the soda to warm up. If instead you have a hot cup of coffee in your room, then heat will flow out of the coffee into the room and the coffee cools down. As the air in the room loses or gains heat, it will also change temperature. You can notice this change by placing your hands near the can or the cup. However, the temperature of the room will change only slightly because the air in it

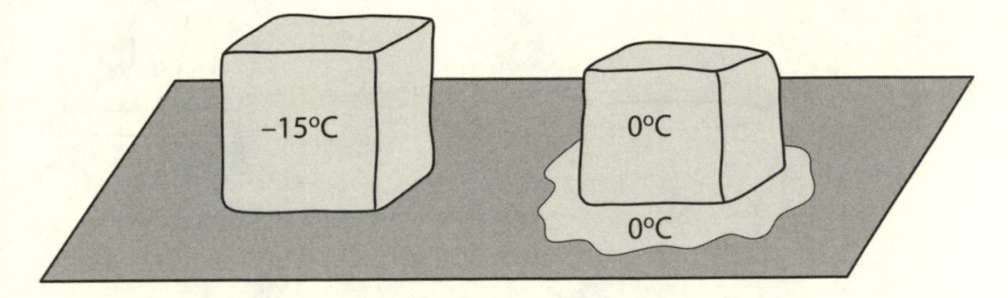

FIGURE 3.5. Heat flow into an ice cube. An ice cube taken from a freezer below the freezing point must first warm up to 0°C. If heat continues to enter the ice cube it will melt, turning into liquid water at 0°C.

is so massive compared to the can or the cup. In addition, there may be a source of heating or cooling that draws warm air or cool air into the room to try to maintain a constant temperature.

When heat flows into or out of an object, there is usually a temperature change, but not always, because some objects may go through a phase change. For example, if you have ice at 0°C and heat is added to it, it begins to melt turning into water at 0°C (fig. 3.5). If you have a pot of water on the stove boiling at 100°C, it remains at that temperature even as heat is continually added to it (fig. 3.6).

If an object goes through both a temperature change and a phase change during a heating or cooling process, the changes occur sequentially one after the other. For example, the freezer in my refrigerator keeps the ice at about –15°C. If I place a piece of ice in the room, heat flows from the room into the ice. If the heat that flows into the ice from the room is distributed equally throughout the piece of ice, then the entire piece of ice will warm up until it reaches 0°C. As heat continues to flow into the ice, it begins to melt. Its temperature remains at 0°C while the ice turns into water at 0°C (again, as long as the heat is distributed equally thoughout) (see fig. 3.5.) Then the water (melted ice) warms up until it reaches room temperature.

A similar situation occurs if I want to cook spaghetti. I fill a pot with water at about 20°C and place it on the stove. The burner gets quite warm, so heat flows from the burner through the bottom of the pot into the water. First, the water heats up from 20°C to 100°C, if the heat is distributed equally in the water. Then it begins to boil and stays at 100°C while boiling (see fig. 3.6). If the steam could be trapped, its temperature

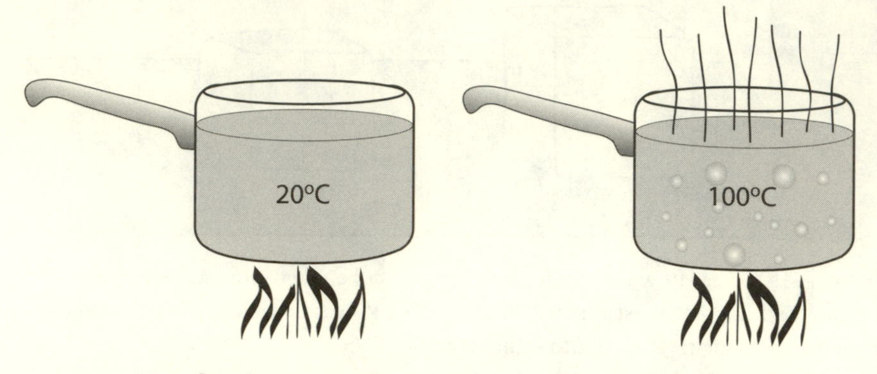

FIGURE 3.6. Heat flow into a pot of water. When heat is added to a pot on the stove, it will heat up until it begins to boil. The water remains at 100°C while the water boils.

would start to increase as more heat is added, but usually it just drifts away. Superheated steam is used in power plants to generate electricity.

What does this have to do with the human body? The human body will not go through a phase change, but only a temperature change. However, other substances that come into contact with the body may go through a phase change. Suppose you sprain your ankle and put an ice pack on it to reduce swelling and pain. Heat from your ankle and foot will flow into the ice pack, causing your ankle to cool and the ice to melt. If you have a covered pot of boiling water on the stove, you may get quite a burn when you remove the lid if the steam comes in contact with your skin.

Another situation that occurs is evaporation of moisture from the skin or the air passages, particularly the oral cavity. Evaporation of water occurs as it changes from a liquid to a gas. While boiling water also changes it from liquid to gas, evaporation occurs at temperatures well below the boiling point. Still, evaporation is a cooling process because the water molecules that leave the liquid phase carry some of their energy with them. This action decreases the internal energy of the remaining water, which causes its temperature to fall. Heat from the body can also supply energy to help in the evaporation process. Perspiration and the resulting evaporation is a very important method of regulating body temperature.

Because heat is a form of energy, there are a number of different energy units that are used. The scientific unit is the joule (J), which makes it easy to compare heat energy with other types of energy and with work.

Other units that are quite common are the calorie (cal) and the kilocalorie (kcal). One kilocalorie is equal to 1000 calories and equal to 4186 J. The food unit of energy is the Calorie, which has a capital "C" instead of the lower case "c" and is not the same as the calorie. It turns out that the food unit Calorie is the same as a kilocalorie, or 1000 calories. It can be a bit confusing, but it is important to understand the differences and how to use them properly. So, to summarize,

$$1 \text{ kcal} = 1000 \text{ cal} = 1 \text{ Cal} = 4186 \text{ J}.$$

A kilocalorie is defined as the amount of heat needed to raise the temperature of 1 kg of water by 1 C° or by 1 K (specifically, from 14.5°C to 15.5°C at 1 atmosphere of pressure). A calorie is defined as the amount of heat needed to raise the temperature of 1 g of water by 1 C° or by 1 K (again, from 14.5°C to 15.5°C at 1 atmosphere of pressure).

Specific Heat and Latent Heat

We can quantify the amount of heat needed to increase or decrease the temperature of a substance or the amount needed to melt or boil a substance by using two parameters: the specific heat and the latent heat. The specific heat c of a substance is defined in terms of the amount of heat Q added to or removed from an object divided by its mass m and divided by its change in temperature, ΔT, or

$$c = \frac{Q}{m\Delta T}, \tag{16}$$

which can be rewritten as

$$Q = mc\Delta T. \tag{17}$$

The specific heat of water is 1 kcal/(kg C°) because if 1 kcal of heat is added to or removed from 1 kg of water, its temperature will change by 1 C°. The average specific heat of human tissue is around 0.84 kcal/(kg C°) although differences occur in fat, muscle, blood, and organs. Specific heats for several types of tissue are listed in table 3.1.[1,2,3] Water has a fairly high specific heat, so that it doesn't warm up or cool off quickly. It takes quite a bit of heat to boil a pot of water on the stove. Hot foods with large water content, such as pizza or a baked potato, contain a large

TABLE 3.1. *Specific heats of human tissue*

Tissue	Specific heat (J/kg K)	Specific heat (kcal/kg C°)
epidermis*	3590	0.86
dermis*	3300	0.79
subcutaneous tissue*	2675	0.64
blood*	3770	0.90
liver†	3620	0.86
lung†	3890	0.93
muscle†	3540 to 3800	0.85 to 0.91
Average	3500	0.84

**Sources:* T. R. Gowrishankar, Donald A. Stewart, Gregory T. Martin, and James C. Weaver. 2004. Transport lattice models of heat transport in skin with spatially heterogeneous, temperature-dependent perfusion. *BioMedical Engineering OnLine* 3: 42. D. A. Torvi and J. D. Dale. 1994. A finite element model of skin subjected to a flash fire. *ASME J. Biomech. Eng.* 116: 250–255.

†*Source:* K. Giering, I. Lamprecht, O. Minet, and A. Handke. 1995. Determination of the specific heat capacity of healthy and tumorous human tissue. *Thermochimica Acta* 251: 199–205.

amount of heat and may burn when you bite into them. The high water content in the body keeps it from cooling or warming quickly.

The latent heat L is defined as the amount of heat added or removed to change the phase of a certain mass of a substance. Mathematically, we express the latent heat as

$$L = \frac{Q}{m} \tag{18}$$

or

$$Q = mL. \tag{19}$$

The term *latent* means hidden, which points to the idea that as heat is added to or removed from an object going through a phase change, there is no change in the temperature. There are two types of latent heat: one referring to the solid-liquid phase change, called the latent heat of fusion, L_f, and the other pertaining to the liquid-vapor phase change, called the latent heat of vaporization, L_v.

The latent heat of fusion of water is 80 kcal/kg and its latent heat of vaporization is 540 kcal/kg. In other words, to freeze 1 kg of water (1 L) at 0°C, 80 kcal of heat must be removed, or to melt 1 kg of ice at 0°C, 80 kcal of heat must be added. Also, 1 kg of steam at 100°C has 540 kcal more heat than 1 kg of (liquid) water at 100°C. This fact helps to explain why it is quite easy to get a severe burn from steam: it contains so much heat.

Now, the human body doesn't normally freeze or boil, even though sometimes we get so mad that it makes our blood boil! However, frostbite, caused by exposure to extreme cold, results in freezing of human tissue. It usually happens at the hands, feet, nose, or ears where circulation of warmer blood becomes restricted. When the body is overheated, say from a fever or overexertion, the body responds by producing perspiration. The amount of heat needed for evaporation is the same as when boiling water turns into steam. Thus, when sweat evaporates a large amount of heat is removed from the body.

Conservation of Heat Energy

Calorimetry is an important topic in the area of heat and thermodynamics. It is a technique in which two or more objects at different temperatures are placed in thermal contact with each other and isolated from the environment. Heat flows from the warmer object(s) into the cooler object(s) until they end up at the same temperature with no net transfer of heat between them (thermal equilibrium). The basic premise is based on the principle of conservation of energy: the heat lost by the warmer object(s) must equal the heat gained by the cooler object(s). It is a fairly simple-sounding concept, but it allows the determination of physical properties about an object, such as mass or temperature or specific heat.

When applied to the human body, calorimetry is a method of keeping track of energy flow within the body or between the body and external objects. Again, I bring up the example of the ice pack applied to a swollen ankle. The conservation of energy principle says that the amount of heat that flows out of the ankle must be equal to the amount of heat that flows into the ice pack. If you are burned by steam from a pot of boiling water, then the amount of heat entering your skin must equal the amount of heat lost by the steam as it cools and/or condenses. The method described here helps to predict how much skin may be affected, or how much ice or steam was involved, so that it can help to determine how long to leave the ice pack in place or give an indication of the seriousness of a burn.

Let's look at an example to see how this method works. Suppose you have a sprained ankle and you are told to treat it by placing an ice pack on it. The ice will cool the tissue including the blood vessels, causing them to constrict and the flow of blood to decrease. The decrease in blood flow will reduce swelling and bruising. Assume that the area to be cooled has a mass of about 0.25 kg and that the ice pack at 0°C cools the area from a normal body temperature of 37°C to around 5°C. Using the specific heat of tissue 0.84 kcal/(kg C°), we can calculate how much ice is needed using the conservation of heat energy:

$$Q_{\text{lost by foot}} = Q_{\text{gained by ice}},$$
$$|mc\Delta T|_{\text{foot}} = |mL_f|_{\text{ice}},$$
$$(0.25\text{ kg})(0.84\text{ kcal/(kg C°)})(37°\text{C} - 5°\text{C}) = m_{\text{ice}}(80\text{ kcal/kg}),$$
$$m_{\text{ice}} = 0.084\text{ kg}.$$

This is 0.084 L or 84 mL, about 3 fluid ounces of ice, or about six ice cubes from my refrigerator. Several assumptions have been made in solving this problem that are not true in real life: the entire area of the ankle does not cool to the same temperature, the surface of the skin may cool closer to 0°C, the inner part will be at a higher temperature, often more ice is used and it doesn't all come into contact with the skin, there is heat that enters the ice pack from the air around it, and so on. However, it gives us a good sense of the numbers that are typical for such an application.

Methods of Heat Flow

We have talked about adding heat to and removing heat from an object. What are the different methods of doing this? There are three methods of heat transfer—*conduction*, *convection*, and *radiation*—and the various ways of heating or cooling an object are based on these three methods. Often there is more than one method of heat transfer at the same time; we will look at some examples.

Conduction is a method of heat transfer where heat flows through an object as a result of the interactions and collisions of atoms and molecules in the object. If one end of an object is at a higher temperature, then the

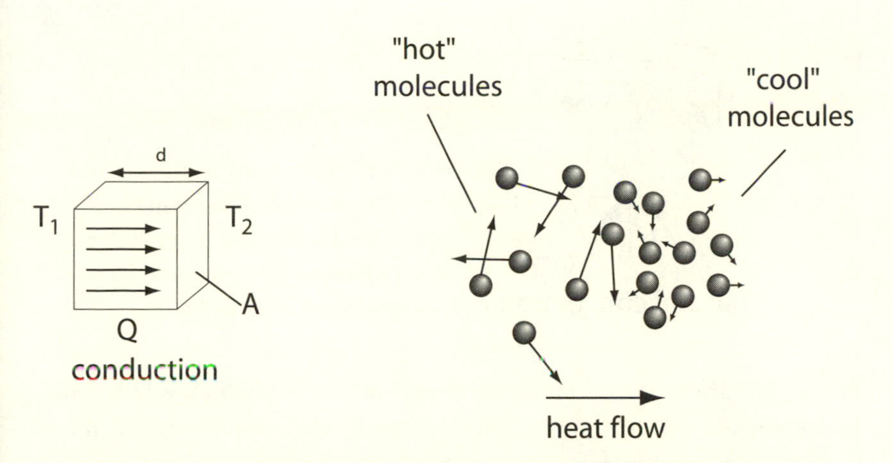

FIGURE 3.7. Conduction of heat. Heat transfer Q by conduction through an object depends on the temperature difference $T_1 - T_2$ across the object, the cross-sectional area A, the thickness d of the object, and the thermal conductivity of the object. When conduction occurs, fast-moving "hot" molecules collide with slower "cool" molecules, giving them more energy. They will then collide with other "cool" molecules, giving them more energy, and so forth, resulting in heat flow through the object.

atoms and molecules at that end move more rapidly, vibrating back and forth about a fairly fixed position, particularly in solids. As these atoms and molecules vibrate, they interact and collide with their neighboring atoms and molecules, giving them more energy and causing their temperature to increase. These now collide with their neighbors, giving them more energy, and the process continues as the heat energy is transferred through the object. Heat will continue to flow at a continuous rate if one end of the object is held at a constant relatively high temperature and the other end is kept at a constant relatively low temperature (fig. 3.7). Conduction occurs in liquids and gases as well, but it is most easily observed in solids, especially metals. An example would be if you were to place a metal spoon in a pot of boiling water—the end sticking out would warm up so that you might burn yourself if you touch it.

Convection is the primary method of heat transfer in fluids (liquids and gases); it involves the transfer of a warmer portion of the fluid to a region of cooler fluid. The result is that "warm" atoms and molecules collide with "cool" atoms and molecules, giving them more energy and raising their temperature. Convection can occur naturally or it can be

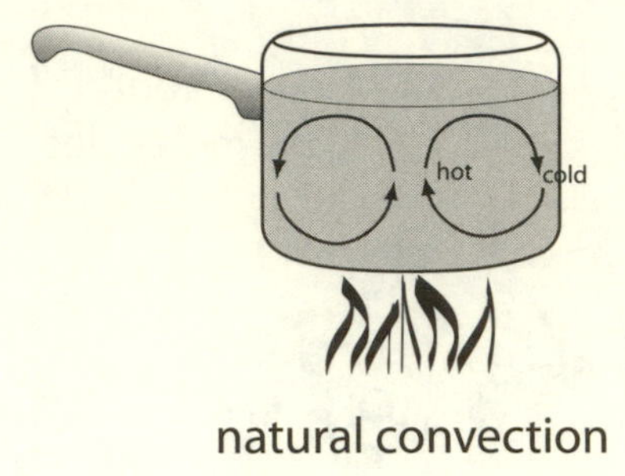

FIGURE 3.8. Convection of heat. Heat flow by convection occurs in fluids (liquids and gases) as warm molecules mix with cool molecules. Convection can occur from natural mixing or it can be forced, as when a fan pushes warm air into a cool room so that it will warm up.

forced. Natural convection occurs when warm fluid, which is less dense, rises because it is more buoyant than cooler fluid. A pot of water will boil because of natural convection (fig. 3.8); an old-style radiator can heat a room when warm air near it rises and mixes with cooler air. If a fan moves warm or cool air around then the convection is forced, as when heating or cooling a house or building, or when cooling an electronic device such as a computer or video projector. Convection ovens are another example: a fan moves the warm air around in the oven.

Radiation is a method of heat transfer that does not have to take place through an object. Radiant energy from the sun moves through the vacuum of space to strike the surface of the earth. Heat is transferred from the sun to the earth without requiring a medium between them (fig. 3.9). You can hold your hands near a light bulb and feel the heat emitted. Cooler objects give off radiant energy as well, but there is a net heat flow from the warmer object to the cooler object.

Often several methods of heat transfer occur for various processes. Let's return to the example of a pot of boiling water. If we focus on the water in the pot, then convection is the main method of heat transfer (although conduction takes place as well). The water near the bottom of the pot warms up, rises due to its decrease in density, and mixes with the cooler water near the top. The cooler water moves downward toward the bottom of the pot, warms up, and then rises to mix with the cooler water. The process continues until the entire pot of water is at 100°C. Then as water at the bottom of the pot continues to gain heat, it turns into steam, which bubbles upward toward the surface.

But how does the bottom of the pot get hot in the first place? The outside bottom of the pot is at a much higher temperature because it is in contact with the burner of the stove. The inside bottom is at the temperature of the water, so heat is transferred through the metal bottom from the burner to the inside bottom of the pot and to the water in contact with it. Conduction in the water causes the water to warm up, which then leads to convection. I remember a few years ago as I was planning to pour the water from a pot of cooked spaghetti noodles into the sink, I reached down to move something in the sink. My arm came in contact with the bottom of the pot and I received a severe burn; I still have the scar today.

With the burner of the stove going or with a roaring fireplace, you can feel heat transfer by all three methods. You can feel the radiation by holding your hands to the side of the heat source. You can feel the warm air (convection) by holding your hands above the heat source. And you can feel the heat through conduction if you place a metal object in contact with the heat source.

Here is a model that helps to quantify heat flow by conduction. The rate of heat conduction Q/t through a material depends on several parameters: the difference in temperature ΔT between the ends of the material, the cross-sectional area A and the distance d of the material through which the heat is flowing, and a quantity called the *thermal conductivity*

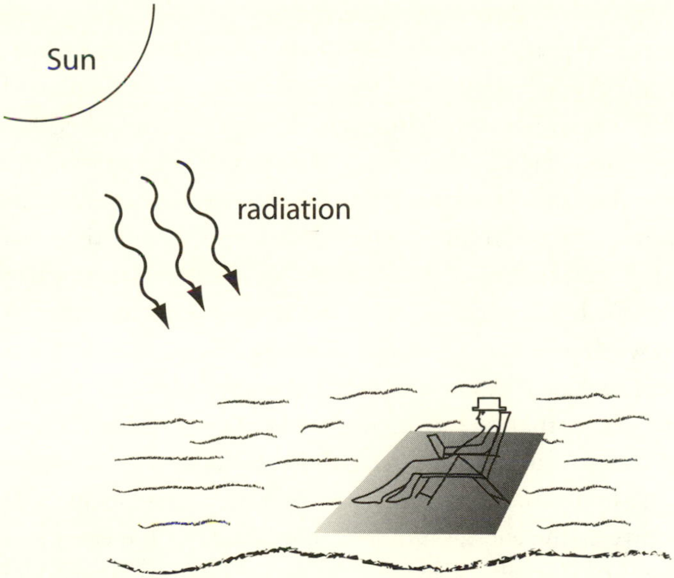

FIGURE 3.9. Radiation of heat. Heat transfer by radiation occurs when electromagnetic radiation is emitted by the hot object and absorbed by the cool object.

given by the parameter k, which depends on the type of material. Mathematically the model is expressed as

$$\frac{Q}{t} = k\frac{A\Delta T}{d}. \quad (20)$$

The model shows that the larger the temperature difference, the greater will be the rate of heat flow (if there is no temperature difference, no heat flows). The model also shows that a larger area and a shorter distance give a larger heat flow. Also, materials that have high thermal conductivity, such as metals, conduct heat very well; insulators have a very low thermal conductivity. We will look at an example in the human body later.

Energy and the Body

We now come to the primary topic of this chapter, which is energy of the human body. We have spent time going through a number of basic physics topics and have related some of them to the human body. As we go through this discussion, we will apply these energy concepts quite often at both the macroscopic and the microscopic levels.

One of the most important physics ideas to keep in mind is the principle of conservation of energy, which states that energy cannot be created or destroyed, but can be converted from one form to another. Another way of stating this principle is that the total energy in a closed system remains constant. And still another way is that any energy that enters or leaves a system must be taken into account when determining the total energy involved. Recall that energy and work are closely related: when an object does work, it loses energy; when work is done on an object, it gains energy. So work plays an important role as well.

When the conservation of energy principle is applied to the human body, the main idea is that energy that goes into the human body is matched by the energy output of the body. Food is the source of energy for the body. The body uses this energy to do work (macroscopic and microscopic), converts the energy into heat, and/or stores the energy as fat (long-term storage) or glycogen (short-term storage) to be used in the cell.

When discussing the energy of the body, its input and output, we won't consider external sources of energy unless we explicitly state so. For example, suppose an elevator lifts a person from the first floor to the fourth floor of a building. In this situation, the elevator is doing work on the person; the person has kinetic energy when moving and gains gravitational potential energy. But this work is external to the body. However, if the person walks up the stairs from the first floor to the fourth floor, then the person is doing work; the body is using energy to perform that work; the energy comes from the food that the person has eaten. Another external source of energy occurs if a person uses a heater to stay warm because heat flows into the body in this situation. Also, we don't consider undigested food because it contributes no energy to the body.

Energy from Food

Food provides the source of energy for our bodies to do work and function properly. The energy from food is potential energy and is stored in chemical bonds. The food unit of energy is the Calorie (large calorie), not the calorie (small calorie). One Calorie is equivalent to 1000 calories or one kilocalorie. The granola bar that I just ate had 130 Cal (130 kcal) according to the wrapper. A large hamburger "with everything" may have over 1000 Cal (1000 kcal), but a diet soda has 0 Cal (0 kcal). I will try to use kcal instead of Cal from now on. By the way, most nutritional values that are stated on food containers and packaging are based on a diet of 2000 Cal per day.

Energy from food comes from three sources, carbohydrates, lipids (fats), and proteins, which are also called nutrients. These three nutrients make up the major part of what we eat when we consume food. Other nutrients include vitamins and minerals, which are needed in only very small quantities, and water, which is important for many different functions of the body. Nutrients in food are used for the body to grow, to be properly maintained, to be repaired, and to do work.

Carbohydrates (sugars, starches, and fiber) are derived almost exclusively from plants. Sugars and starches are digested to form glucose, which is used by cells to produce adenosine triphosphate (ATP). ATP is a chemical that releases its energy to be used by cells in the body. Fiber comes in two forms, one which is insoluble to help with eliminating waste and the other soluble to help reduce blood cholesterol. One gram of carbohydrates produces about 4 kcal of energy in the body.

Lipids or fats have several uses in the body, which include helping the body absorb and store vitamins (A, D, E, and K), providing protection to organs, serving as a heat insulator, and providing energy. Triglycerides are used as a source of energy and include saturated fats from animal products and unsaturated fats from plant products. About 9 kcal of energy is produced per gram of lipids.

Proteins are important components of tissue (muscle, skin, etc.) in the body and are needed for growth and repair of the body. Although carbohydrates and lipids are the main sources of energy, proteins do provide some energy. One gram of protein provides about 4 kcal of energy, which is the same as for carbohydrates.

Energy from carbohydrates is delivered most quickly, and that from fats most slowly. It is recommended by food nutritionists and scientists that about 50 to 60% of a person's calories come from carbohydrates, about 20 to 30% from fats, and about 15 to 20% from protein.

Work Done by the Body

When a person does work, they must use energy. When you pick up your backpack, walk across the room, or climb a flight of stairs, you are doing work. These are examples of work on a macroscopic level, external to the body. But work also goes on at the microscopic level as forces push blood, cells, and molecules in the body. Also, work is done to move arms and legs and other parts of the body.

Work done at the macroscopic level actually isn't all that impressive. Let's look at a quick calculation to understand why. Suppose a 70 kg person (about 150 lbs) walks up the stairs to go from the first floor to the second floor, a vertical distance of 3 m (about 10 ft). The amount of work done by the person turns out to be the potential energy at the second floor relative to the first floor:

$$W = U = mgh = 70 \text{ kg} \times 9.8 \text{ m/s}^2 \times 3 \text{ m} \approx 2060 \text{ J} \approx 0.5 \text{ kcal}.$$

In other words, the person expends only half a kilocalorie of energy (fig. 3.10). It doesn't seem like much. According to this calculation, if you

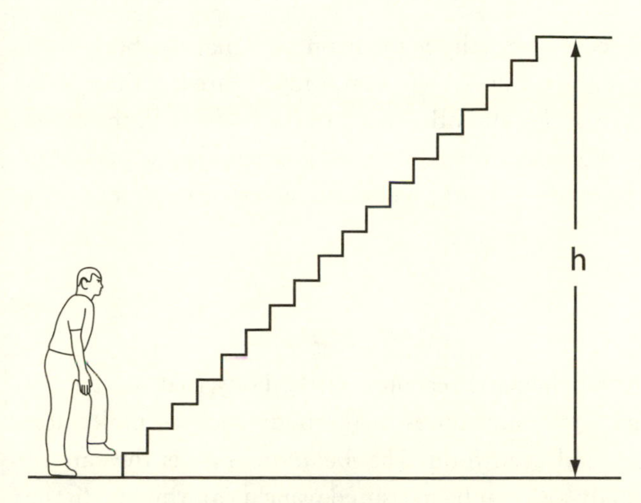

FIGURE 3.10. Work. The work done by a person climbing a flight of stairs depends on the body mass and the vertical distance *h* moved. However, work done by the body to accomplish this task also includes moving the arms and legs, breathing, and pumping blood through the body.

want to work off the 100 kcal snack pack that you just ate, you need to walk up 200 floors!

Lifting weights, running, climbing stairs, and riding a bicycle are all examples of exercises used by people to burn calories. These exercises do not require a lot of energy to complete in terms of the mechanical work done at the macroscopic level. However, charts that show the energy used to do such things indicate that the amount of energy is much higher. For example, about 400 to 600 Calories per hour are burned during (machine) stair climbing. The amount depends on how vigorous the exercise is: Is it continuous? Is it hard, going all out? Is it slow and easy? To burn the 100 Calories in the snack pack, you would need to walk up stairs for about 10 to 15 minutes according to the charts. If it takes about 15 seconds to walk up one floor, then you would have to walk up 40 to 60 floors to burn that 100 Cal, which is very different from 200 floors.

So something else must be going on, and indeed that is the case. To go up the stairs, you must swing your arms and legs, working them vigorously as well. Also, work is being done internally at the microscopic level to move blood through your body, air into and out of the lungs, particles into and out of cells, and to break down nutrients in your body.

The work that occurs internally is the main way that the body uses energy, and it results in the production of heat. Muscles are less than 20% efficient in moving arms, legs, and other parts of the body, with the other 80% or so of energy being converted into heat. Even when a person is resting, not doing any external work, the body converts energy into heat continuously.

Metabolism

Metabolism refers to biochemical reactions in the body that use energy to build up and tear down substances in the body such as molecules, cells, proteins, tissue, and even food. The *metabolic rate* is the rate of energy usage in the body and can be measured using a calorimeter. In the calorimeter, a person is submerged in water and the temperature rise of the water is measured. Using conservation of heat energy, the amount of heat leaving the body must equal the amount of heat gained by the water. From this information the metabolic rate can be determined.

The rate at which the body converts energy into heat while resting is called the *basal metabolic rate*, or BMR. This resting rate is required for a person to breathe and for blood to pump through the body; slight differences occur if digestion of food occurs or not. For an adult male whose mass is 70 kg, the BMR is about 70 kcal/h. For a 70 kg adult female, the BMR is about 60 kcal/h.[4] This means that, over the course of a 24 h period, the male burns about 1700 kcal and the female about 1500 kcal without doing any external work. There are many variables that affect a person's BMR, including gender, age, and the surface area of the body. Men have higher BMR because they have more muscle. Children and teenagers have higher BMR because they need more energy to grow. Tall, thin people have higher BMR because they have more surface area per mass.

Total metabolic rate (TMR) takes into account the total energy used by the body to perform all of its functions, including internal and external work. When a person exercises, the TMR can go up quite a lot. If the BMR means that a person is burning 1700 kcal per day, when the various activities of the day are included, the actual requirement may be 2500 kcal for the day. The external work, as we have seen, may not be substantial at all; the real increase occurs for internal work, that is, to pump blood, to breathe, to digest food, to maintain a constant body temperature, and to move parts of the body. If a typical person burns 500 kcal/h to climb stairs, about 70 kcal/h would be burned even if the person were resting, and the other energy is burned because of the increased activity.

Heat and Body Temperature

If so much of the energy that a person consumes is converted into heat, what happens to that heat? Doesn't a person's body temperature stay fairly constant? If heat is continually added to the body, shouldn't it heat up continually? Why would we ever get cold? The answers to these questions come from an understanding of how the body regulates its temperature through the removal of heat from the body, as well as the addition of heat to the body.

An important function of the body is called *homeostasis*. Homeostasis is a property in which internal conditions of the body remain relatively stable over time while conditions outside the body may be in a constant state of change. The temperature of the body is very stable around 37°C (98.6°F), fluctuating by only few degrees during exercise or illness. Actually, this value is the core body temperature associated with organs in the skull, chest, or abdomen. The temperature of the shell, which is primarily the skin, varies quite a lot depending on external temperature. The temperature external to the body can change dramatically from a cold winter to a hot summer, but the temperature of the core of the body stays nearly constant. If a person dies and internal processes cease, heat is no longer produced and the temperature of the body changes; this change in temperature can be used to determine the time of death. Episodes of the crime drama *CSI* (Crime Scene Investigation) use this technique often.

The ability of the body's temperature to remain constant is very important for many processes that occur in the body. This regulation of temperature occurs when there is a balance between the production and the loss of heat. The main sources of heat are the metabolic processes in the major organs of the body, such as the liver, heart, and brain, as well as skeletal muscles. Movement of skeletal muscles through physical activity can cause a dramatic increase in the amount of heat produced. Heat is lost primarily through the skin in a variety of ways, including conduction, convection, and radiation, as well as evaporation. Usually there is a combination of methods of heat loss involved.

Let's look at heat loss by the body without considering any external activities initially. We assume that the outside air temperature is lower than normal body temperature. First, heat loss by conduction occurs when internal organs and circulating blood are at a higher temperature and the skin is at a lower temperature. Second, heat loss by convection occurs when the cooler air in contact with the skin is heated up; then it rises, moving away from the skin, and is replaced by cooler air. Third, heat loss by radiation (in the infrared region) occurs because any warmer object radiates heat. And last, evaporation occurs through the skin, the mouth, and the lungs as water changes from liquid to gas, not because it is boiling, but because the water molecules in contact with these tissues gain enough kinetic energy (vibrational energy) to escape the liquid phase and turn into water vapor. Of all the heat lost, conduction and convection make up about 20%, radiation about 50%, and evaporation the remainder.

Under typical conditions, the body loses heat at the rate of roughly 100 J/s, or 100 W. This means that the body gives off about the same amount of heat as a 100 W light bulb (fig. 3.11). In order to properly control the temperature in buildings, the number of occupants should be taken into account as a heat source in designing heating and cooling systems.

The rate of heat loss (or gain) by all of these methods is affected by many conditions. A person can lose heat by swimming in cool water, drinking cold water, or breathing cold air. Heat gain can occur during a hot shower or bath. The rate of heat exchange by convection can be dramatically different if a person is clothed or if bare skin is exposed. Wearing a jacket on a cold day helps to trap air warmed at the skin. Cooling down by standing in front of a fan that blows cool air is called *forced* convection. Radiant energy from the sun can warm a person if it strikes the skin, as when sunbathing. Heat loss through evaporation increases when the humidity is low, or when water is splashed on the skin and air is blown across it. The heat index and wind-chill factor are tools used to indicate the body's effective heat loss and/or gain based on atmospheric conditions.

So if the body can generate, gain, and lose heat, how does it go about regulating its temperature? Regions of the brain, particularly the hypothalamus, receive signals from heat receptors in the core and in the shell. If the body temperature is too low, then these signals tell the body to

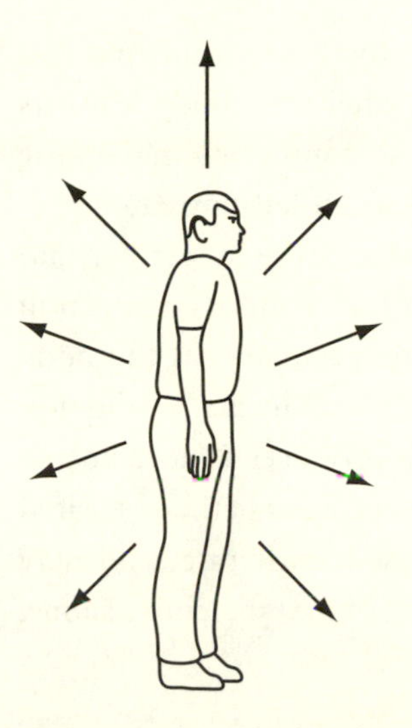

FIGURE 3.11. Heat loss. A typical person gives off about 100 J of heat per second, which is equivalent to the rate of energy production of a 100 W light bulb.

increase the amount of heat generated or to decrease the amount of heat lost. If the body temperature is too high, these signals tell the body to increase its ability to lose heat.

There are several actions that the body can take to increase the amount of heat generated or decrease the amount of heat lost. One method causes the blood vessels of the skin to be constricted, which decreases the flow of warm blood to the shell, resulting in a decreased loss of heat by conduction, convection, and radiation. Another method is shivering where the brain causes skeletal muscles to move involuntarily, accompanied by the production of heat. And still another method involves the release of certain chemicals that cause the metabolic rate to increase. These methods are all involuntary. Voluntary actions include wearing clothing, putting on a blanket, drinking hot fluids, taking a warm bath, and increasing physical activity.

If body temperature is too high, there are several ways to enhance heat loss. The body may cause blood vessels near the skin to dilate, which increases blood flow, resulting in more heat flowing from the core to the shell so that heat can be dissipated. The body may also produce larger quantities of perspiration than normal, so that evaporation is increased. Taking a cool bath or shower will help to cool the body as well.

Hyperthermia and *hypothermia*, elevated body temperature and low body temperature, respectively, are conditions where the body loses its ability to lose or produce heat as needed. These conditions may cause significant health problems that could lead to death if left untreated.

You may have experienced elevated body temperature (hyperthermia) when you have been ill, resulting in a fever. Fever is a result of an infection or other serious illness or injury. Heat loss is reduced so that the body temperature increases. This increase is thought to aid in the production of white blood cells that kill infected cells, and it produces a more hostile environment that helps to destroy bacteria. Fever is sometimes treated because it can be uncomfortable, causes increased heart rate, and may result in dehydration. If the fever is too high there may be convulsions, hallucinations, and even brain damage.

Specific Heat and Thermal Conductivity of the Body

Take the example of someone who has a high fever and needs to be cooled down to avoid going into convulsions or suffering brain damage. Suppose a 70 kg person with a temperature of 40°C (104°F) is placed in a tub filled with cool water at 20°C so that the body temperature can be lowered to normal (37°C). The amount of heat that needs to be removed is calculated from the expression $Q = mc\Delta T$, where the average specific heat of the human body is 0.84 kcal/(kg C°) or 3500 J/(kg C°):

$$Q = mc\Delta T,$$
$$Q = (70 \text{ kg})[0.84 \text{ kcal/(kg C°)}](40°\text{C} - 37°\text{C}),$$
$$Q = 176 \text{ kcal} \approx 740{,}000 \text{ J}.$$

In this problem, the removal of the heat will occur primarily through conduction. The mathematical model [eq. (20)] proposed earlier for the rate of conduction is given by

$$\frac{Q}{t} = k\frac{A\Delta T}{d}.$$

For the human body, the average thermal conductivity is 0.20 J/(m s C°) and the average area of the skin for a 70 kg person is 1.8 m^2.[5,6] Conduction will typically occur through the outer part of the shell, with the inside at a higher temperature and the outside at a lower temperature. For our purposes, we will assume that the thickness of tissue through which the heat flows is about 1 cm or 0.01 m. When we plug these numbers into the expression above we find

$$\frac{Q}{t} = 0.20\frac{(1.8)(20)}{0.01} = 720 \text{ W}.$$

If the total amount of heat needing to be removed is 740,000 J and it is removed at the rate of 720 W, then it will take over 1000 seconds or about 17 minutes to lower the body temperature.

Now I concede that there are difficulties with the assumptions here: the person will continue to produce heat; the average thicknesses, areas, and other values may be different; the water temperature will likely increase; and so on. But this calculation gives us a number that seems to make sense. If a person with a high fever is placed in a cool tub of water, you would expect that the temperature would lower somewhat within 15 to 20 minutes.

Energy Imbalance

Food is the primary source of energy for the body. Energy is used and lost in doing work and generating heat. If the energy available from food matches the output of energy through work and heat loss, the weight of the person remains quite stable. If, however, there is an imbalance, weight gain or loss may occur.

You may realize through your own experience that for most people it is easier to gain weight than to lose it. Eating and drinking foods that are high in carbohydrates and fats can lead to weight gain and perhaps obesity—the excess caloric content is converted into fat that is stored in the body. With increasing age, a person's metabolic rate may decrease so that less energy is used than at a younger age.

However, gaining weight is not always bad. Weight gain in children is important as they grow and change. Pregnant women will gain weight as the unborn baby develops and grows. Some people who are underweight can gain weight through diet, as well as exercise such as weightlifting, which helps to increase the size and mass of muscle tissue.

Weight loss may be necessary to promote good health, or it may result from a serious illness. Intentional weight loss is best accomplished by limiting calories (dieting) and increasing physical activities. There are many diet plans that have some variation or combination of limiting carbohydrates and fats. Other methods to lose weight include medication to decrease hunger or to increase metabolism. More extreme methods include surgical procedures to reduce the size of the stomach or to remove fat (liposuction).

Summary

I have tried to give you some understanding about work, energy, and heat as it applies to the human body. What goes on at the molecular level in breaking down large molecules of carbohydrates, fats, and proteins is the subject of entire courses in biochemistry and nutrition. The areas of exercise, weight loss, and overall health can be studied in much more depth as well. I hope that you can begin to explore these concepts in more detail through this discussion of some of the basic ideas.

FOUR

Sound, Speech, and Hearing

Do you like to sing, whether in a choir, a rock band, or the shower? Do you play a musical instrument, such as trumpet, flute, guitar, or violin? Do you like to listen to music on the radio, at a concert, or on your MP3 player? Why does a flute sound different from a trombone, or a violin from a guitar? Why do people lose their ability to hear higher frequencies as they age?

Many years ago a relative told me that she thought the most important advancement in the future would be "communication." I didn't really understand what she meant at the time (there was a failure to communicate) because I thought she was talking about technological advances, and I couldn't see that anything could be better than a CB (citizen's band) radio.

While there are many ways to communicate, using cell phones, the internet, newspapers, and even nonverbal gestures, one of the most important senses of the human body is hearing or the ability of the human body to detect sound. In this chapter we will look at basic properties of sound and describe how humans hear. We will also look at ways of producing sound by the human voice and other devices, and explore topics such as loudness, earplugs, ultrasound imaging, and cochlear implants.

Waves

Sound is an example of a wave. Waves are disturbances that move through a region of space. There are lots of examples of waves. A water wave

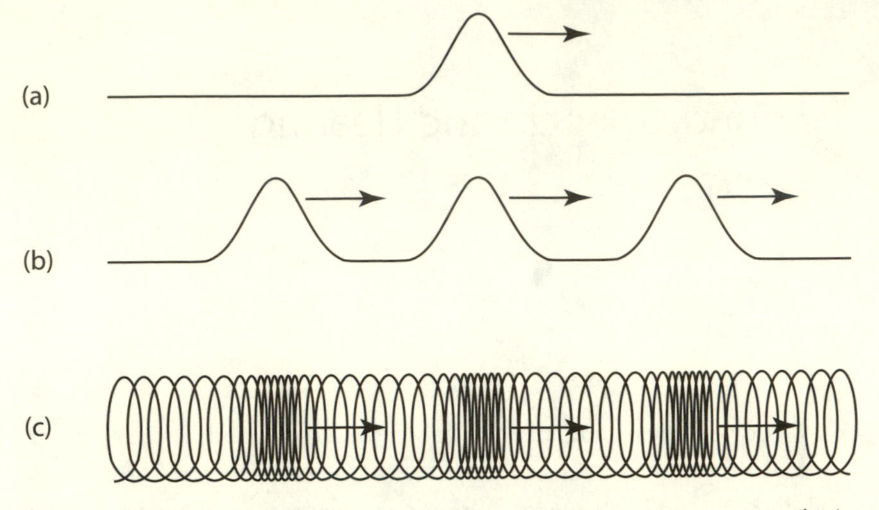

FIGURE 4.1. Wave motion. (a) A single wave moving along a rope or string. (b) A periodic wave consisting of a series of waves at regular intervals moving along a rope or string. Both (a) and (b) are transverse waves because the particles that make up the rope move perpendicular (up and down) to the direction in which the wave is traveling (to the right). (c) A periodic wave moving through a spring. This wave is a longitudinal wave because the particles in the spring move parallel (left and right) to the direction in which the wave is traveling (to the right).

crashes on the beach at the ocean or forms when you toss a rock into a calm lake. A wave is easily produced in a rope or string if you quickly shake it up and down. A slinky has a wave in it when you move it back and forth from one hand to the other.

A wave can be a single disturbance that occurs at one point in time or a series of disturbances that occur in a repetitive way; these are called *periodic waves* (fig. 4.1). Usually many water waves crash on the beach, one after the other in a periodic fashion. If you drop several pebbles into the lake, one by one at regular intervals, you will produce a periodic wave. If you shake a rope up and down, up and down, you can form a series of waves in the rope. If the timing is just right as you shake the rope, the waves will vibrate in segments so that they do not appear to be traveling along the rope, but staying in one place. This action results in a *standing wave*. One of the simplest standing waves occurs when two people use a jump rope by each holding it at one end and moving it up and down, or round and round.

Sound waves are periodic waves. When a guitar string is plucked it vibrates back and forth repetitively for several seconds. The vibrating guitar string is not a single wave, but a periodic wave; it loses energy after a few seconds and has to be plucked once again to keep vibrating. The sound produced by the vibrating guitar string is periodic as well; the loudness diminishes over the time that the string is vibrating, but returns once it is plucked again. If you clap your hands, or strike a nail with a hammer, or drop an object on the floor, the sound that is produced may seem like a single wave, but it is actually a complex periodic wave that lasts only a short time.

As a wave passes through a certain region of space, the material (or medium) that the wave is traveling through physically moves. (Electromagnetic waves, such as visible light, can move through a vacuum, but for this chapter we will not be discussing them.) Sometimes the medium is moving with the wave, as when a water wave crashes on the beach. Most of the time the medium moves a short distance, and then returns to its original location after the wave passes. The wave continues to move through the medium, but the material or medium does not move with it. An example is the water wave produced by dropping pebbles. If you were to observe a bug or some other floating object on the surface of the water, you would see it bob up and down, but stay generally in the same location. Another example is the rope that you shake very quickly—the wave moves along the length of the rope, but the particles that make up the rope do not.

A wave is either *transverse* or *longitudinal* based on the direction of motion of the medium. Suppose a wave is moving from left to right in front of you at this moment. If the material moves up and down (perpendicular to the direction of the wave), then the wave is transverse. If the material moves left and right (parallel to the direction of the wave), then the wave is longitudinal. Jump ropes, guitar strings, and water waves are all examples of transverse waves (fig. 4.1b). The slinky is an example of a longitudinal wave (fig. 4.1c). Sound is also a longitudinal wave that we will describe in more detail later.

Think about watching the "Wave" at a football game. To start the wave a few people in a certain section of the stadium stand up and then sit down. If a few more people next to them stand up and sit down a few moments later, and the process continues around the stadium, you have just observed a transverse wave. The crowd is the medium and the individual people are the particles of the medium. As the motion of the Wave travels sideways, the particles move up and down, perpendicular to the direction of travel of the wave.

Now consider a variation on the Wave. Suppose the people who started the wave were to move to the right until they bumped the next person with their shoulder, and then returned to their upright position. Then the next person does the same thing, and the next, and the next, as it continues around the stadium. This would be an example of a longitudinal wave: the particles are moving left and right, as the Wave makes its way sideways around the stadium.

Simple Harmonic Motion

Periodic waves are produced by or in objects that vibrate repeatedly. They usually follow the rules of what is called *simple harmonic motion* (SHM). When SHM occurs, the motion of the vibrating object is described by several properties. First, when the object is at rest before it starts vibrating, it is in its equilibrium position. When it oscillates, it moves between two locations on either side of the equilibrium position. The maximum movement of the vibrating object is called the *amplitude* of the vibration. If there is no loss of energy, the object will continue to vibrate with no change in the amplitude.

Second, the object will go through many cycles of motion over a period of time. The number of oscillations per unit time is called the *frequency*, and the amount of time for one oscillation is called the *period*. The period T is the reciprocal of the frequency f, or

$$T = \frac{1}{f}, \quad (1)$$

and the frequency is the reciprocal of the period. Thus, for high frequencies, the period is small, which means that the object vibrates back and forth very rapidly. At low frequencies, the object takes a longer time to vibrate back and forth. Frequency is usually measured in hertz (Hz); the hertz is a reciprocal second, or s^{-1}. That means that the period is measured in seconds.

For sound waves, the amplitude is related to the loudness and the frequency is related to the pitch. If you pluck a guitar string lightly, it will produce a soft sound, but if you pluck it quite hard, it will be much louder. The loudness of the sound produced in each case is directly related to the amplitude of the vibrating string. If you pluck one of the thick strings on the guitar, then the frequency is lower than the frequency of one of the thin strings. The sound produced by the thick string has a lower note, or pitch, than for the thin string.

A periodic wave that moves through a material has a speed that depends on properties of the material, the frequency or period, which depends on how fast the material oscillates, and another characteristic called the *wavelength*. The wavelength is the distance from the crest of one wave to the crest of the next wave, such as between peaks of water waves crashing on the beach, or between peaks of waves in a rope.

These quantities (speed, frequency or period, and wavelength) are related: if the wave goes through one complete oscillation in a time equal to one period, then it will move through a distance of one wavelength, which depends on how fast it is moving. Using the simple expression that speed is equal to distance divided by time, it is easy to show that the speed v of the wave equals the wavelength λ divided by the period T (or times the frequency f). In symbol form, this relationship is usually written as

$$v = f\lambda. \tag{2}$$

Again, the speed depends on the properties of the medium and is constant as long as the medium doesn't change. Thus, the relationship shows that, as the frequency of a wave increases, its wavelength should decrease.

For sound waves moving through air, the speed of sound is fairly constant at around 340 m/s. Thus for low frequencies the wavelength is larger, whereas for high frequencies the wavelength is shorter. Often the size of the device used to produce the sound is related to how low or high the frequency will be. A large musical instrument like a tuba produces low-frequency sounds, whereas smaller instruments like the piccolo or flute produce high frequencies. Another example is the piano: the lower notes are produced by longer strings and higher notes are made by shorter strings.

However, as with most things the relationship between size and frequency is not quite so simple because the speed of a wave in the object depends on other properties of the object. For example, in the piano the frequency of each string depends on how tightly it is stretched and how thick or massive it is. We will see this relationship in the human vocal cords as well.

To understand this dependence, we take two examples: the vibrating spring and the vibrating string. A certain amount of force is needed to stretch a simple spring. If a mass is attached to the spring, the mass-spring system will vibrate with a frequency given by the expression

$$f = \frac{1}{2\pi}\sqrt{\frac{k}{m}}, \tag{3}$$

where k is the spring constant and m is the mass of the object attached to the spring. The spring constant is related to the stiffness of the spring. A slinky does not require a lot of force to stretch it, so it is not very stiff. Thus, it has a small spring constant. However, a car spring, which is attached to the axle of each tire, is a very stiff spring because it requires a large force to stretch or compress it. Thus, it has a large spring constant. According to the frequency formula above, we see that for a large spring constant and small mass, the frequency is high, whereas for a small spring constant and a large mass, the frequency is low.

For the vibrating string, its fundamental, or lowest, frequency is given by the expression

$$f = \frac{1}{2L}\sqrt{\frac{T}{\mu}}, \tag{4}$$

where L is the length of the string, T is the tension in the string, and μ is the linear density (mass per length) of the string. From this expression we see that the shorter the string the higher the frequency, the tighter the string the higher the frequency, and the thinner the string the higher the frequency. If you look at the strings on a piano, you will see that the higher notes are produced by shorter, tighter, thinner strings and the lower notes are produced by longer, looser, thicker strings. The strings on a guitar or violin all have the same length, but the lower notes are produced by thicker strings and higher notes by thinner strings. You may also have experience in trying to tune a guitar or a violin by tightening or loosening the strings to change the frequency.

We have looked at why sound is considered a wave, and what it means to be a periodic wave. Now, why is sound a longitudinal wave?

Pressure Wave

In saying that sound is a longitudinal wave, we need to recall that for longitudinal waves the particles that make up the medium move parallel to the direction of the sound wave (see fig. 4.1c). Another name for a sound wave when it is in a gas, such as air, is a pressure wave. This term applies because, when the molecules of the gas move back and forth, sometimes they are closer together and other times they are farther apart. The molecules that move closer together will cause an increase in the number of molecules in a certain region of space. When this happens, the pressure increases as it does when air is added to a tire. If the molecules move farther apart than normal, then the pressure decreases. Thus there are alternating regions of high pressure, called *condensations*, and regions of low pressure, called *rarefactions*, which are slightly above and below the normal atmospheric pressure of the air in which the wave is moving (fig. 4.2). This increase and decrease of pressure is a simple application of the ideal gas law discussed in chapter 2 and given by the equation

$$pV = Nk_BT. \tag{5}$$

If the number of molecules N increases in a certain amount of space given by the volume V, then the pressure in that space increases. If the number of molecules decreases, then the pressure decreases as well.

But sound waves are not restricted to gases. They can travel through liquids and solids too. If you have ever been swimming, perhaps you have heard a friend try to talk to you when underwater. The sound is quite odd, but can be understood. Submarines and ships use sound waves to determine if submerged objects are nearby. Sound travels through solids as well. Perhaps you have made a "telephone" out of two cups connected by a string. If a friend talks into one cup, you can hear what they are saying

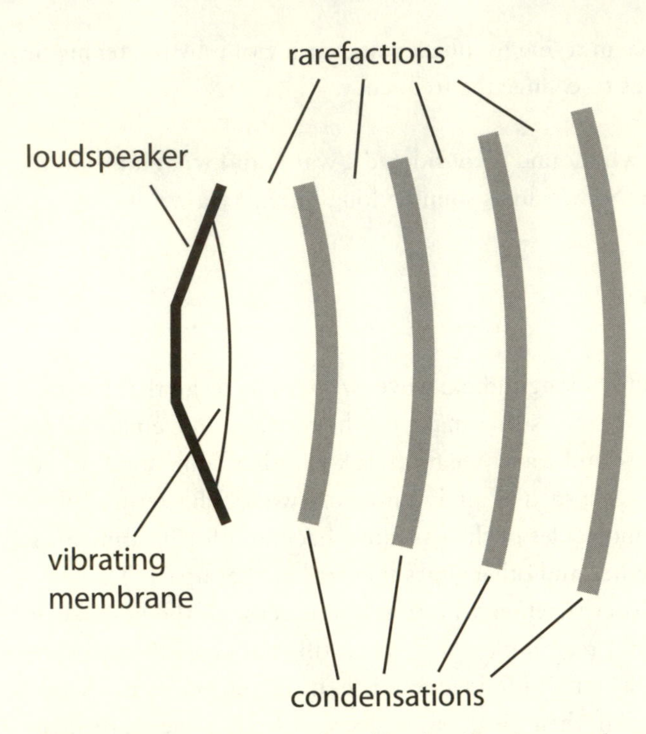

FIGURE 4.2. Sound wave produced by the vibrating membrane of a loudspeaker. Regions of relatively high pressure are called condensations and regions of low pressure are called rarefactions.

if you hold the other cup up to your ear. Real telephones (landlines) don't transmit sound waves however; sound waves are converted into electrical signals by the telephone of the person who is talking; these electrical signals are transmitted through the wires to the other telephone, which converts it back into a sound wave for the person listening.

As a sound wave travels through a liquid or a solid, the molecules and atoms move back and forth parallel to the direction in which the sound wave is traveling, because it is still a longitudinal wave. However, it doesn't really make sense to say that the sound wave is a pressure wave in liquids and solids. The molecules and atoms return to their normal positions quickly because of the strong bonds between them.

Because sound waves involve the motion of atoms and molecules, sound cannot travel through a vacuum. There must be a medium for the sound wave to pass through. If there are no atoms or molecules in some region of space, there can be no pressure wave produced. Even a vacuum is not perfect and still contains a small number of molecules. However, there is not a large enough change in pressure of the few molecules that remain for the sound wave to move.

The Speed of Sound

Sound waves travel at different speeds in different materials. One of the most important factors determining the speed of sound is the elasticity of the medium. Solids tend to be quite elastic, so sound travels fastest in solids. Steel is highly elastic, with one of the largest values of the speed of sound: about 5100 m/s, or 11,400 mph. Liquids are not as elastic as solids, so the speed of sound is smaller. The speed of sound in water is about 1500 m/s, or 3400 mph.

In gases at atmospheric pressure, the speed of sound varies depending on the mass of the molecules and the temperature of the gas. Gas molecules that have smaller mass respond more quickly (accelerate) when a force is applied to them; lighter gases such as helium have higher speeds than heavier ones such as air (consisting of mostly nitrogen and oxygen). At 0°C, the speed of sound in helium is 972 m/s (2182 mph) compared to a speed in air at 0°C of 331 m/s (743 mph). At higher temperatures, gas molecules are moving faster, which correlates with a higher speed of sound. Thus, for a given gas, as the temperature increases the speed of sound increases. In air, the speed of sound is 331 m/s (743 mph) at 0°C and 343 m/s (770 mph) at 20°C.

Travel faster than the speed of sound (in air) by humans was a goal in the mid-1900s. The first aircraft to officially exceed the speed of sound was piloted by Chuck Yeager on October 14, 1947. Breaking the sound barrier was a routine event by the late 1950s. When I was a kid in the 1960s, I lived about 30 miles from an air force base, and the flight path of the jets was often near my home town. I vividly recall hearing the sonic booms occurring on a regular basis as airplanes passed overhead.

Breaking the sound barrier on land was first achieved on October 13, 1997, almost exactly 50 years after Chuck Yeager's flight. A British team running their jet-powered car (the ThrustSSC) in the Black Rock Desert of Nevada broke the sound barrier twice. In order to set the official supersonic land speed record, the team had to break the sound barrier by running the car first in one direction and then in the other within one hour of each other; however, these two runs were separated by 60 minutes and 50 seconds, just outside the rules of the competition. But two days later on October 15, the team established the record when their two runs were separated by less than one hour. They were measured at 759.333 mph and 766.609 mph, both just over the speed of sound![1]

Production of Sound

You may recall that I mentioned previously that *objects produce sound waves*, as when a guitar string is plucked or a tuba is played or a nail is struck by a hammer. How is this sound produced?

Sound waves in gases are also called pressure waves. Regions where the air molecules are closely spaced are condensations and regions where they are farther apart are rarefactions. When an object, such as a guitar string, vibrates, it repeatedly pushes and pulls on the air around it. As it pushes on the air, the pressure increases, and as it pulls on the air, the pressure decreases. This principle is basically all there is to production of a sound wave. If an object is mechanically moving, it pushes and pulls on the air near it and a pressure wave or sound wave is produced.

As a guitar string vibrates, it pushes and pulls on the air around it in different directions so the sound waves move outward from the guitar in all directions. This is a point source of sound, where the sound moves away from the source equally in all directions. (We will discuss point sources in more detail later.) If the guitar string vibrates in its fundamental (or lowest) frequency, then it produces a sound wave with that frequency. Figure 4.3a shows a *simple* periodic sound wave produced by a vibrating object like a guitar string. Usually the entire guitar, including the sound board, vibrates with a mixture of frequencies, so the sound that the guitar produces consists of many frequencies. The resulting sound wave is called a *complex wave* (fig. 4.3b).

A loudspeaker from a stereo system is more complex than a guitar string, but the principle is pretty much the same. A membrane, called a diaphragm, consisting of cardboard or plastic is stretched across a circular ring or tube and is attached to a magnet. The magnet is moved by

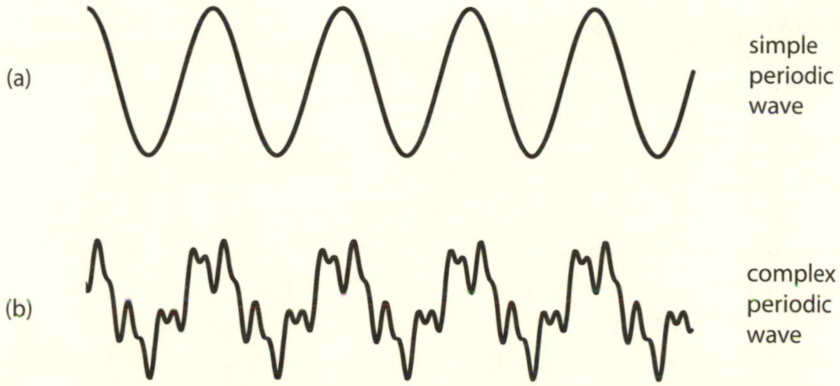

(a) simple periodic wave

(b) complex periodic wave

FIGURE 4.3. Periodic waves. (a) A simple periodic sound wave has a single frequency. It is a pure tone that is produced by a tuning fork or a vibrating string, such as a guitar string or a piano string. It takes the form of a sine or cosine wave. (b) A complex periodic wave occurs when other frequencies mix with the fundamental frequency, as when the entire guitar vibrates or when a person sings a certain note.

electromagnetic forces produced by a coil of wire and causes the diaphragm to vibrate. The diaphragm pushes and pulls on the air in front of it, producing a sound wave. Different parts of the speaker have different sizes and thicknesses so that they vibrate at different frequencies. The larger parts of the speaker produce the lower notes and the smaller parts produce the higher notes. Large speakers, called woofers, are best for producing high-quality sound at low frequencies; small speakers, called tweeters, are used for producing high frequencies. The loudspeaker is not a point source of sound; it is more unidirectional because the sound waves move with larger intensity through the region of space directly in front of the speaker.

The sound that occurs when a hammer strikes a nail is even more complex. However, the same basic principle applies as the hammer-nail system vibrates, pushing and pulling on the air around it, producing a pressure wave. For the hammer-nail system, the vibration is over a fairly short time, but it turns out to be quite long compared to the period of the wave produced. For example, the most common frequencies for human hearing are between 100 and 5000 Hz. These frequencies correspond to periods ranging from 10 to 0.2 ms. If the hammer striking the nail is in contact for a time of 100 to 200 ms, the sound produced by the hammer-nail system will consist of many wavelengths. When these frequencies are mixed together, they produce a complex wave.

Detection of Sound

The detection of sound is, in its most basic form, the opposite of the production of sound. When a sound wave, or pressure wave, strikes an object, it repeatedly pushes and pulls on the object, causing it to vibrate. If this vibration can be measured, the sound wave can be detected.

Sound waves are detected primarily in two ways, by a microphone or by the human ear. A simple microphone consists of a plastic or paper diaphragm, or membrane, stretched across a circular ring or tube attached to a magnet that moves through a coil of wire (sound familiar?). Electrical signals are produced and are detected by electronic equipment. The electronic equipment either stores the signals or converts them for use in a loudspeaker. Because microphones and loudspeakers are so similar, a microphone can be used as a loudspeaker and a loudspeaker can be used as a microphone, although the quality of sound is greatly diminished.

The eardrum is basically a membrane or diaphragm that vibrates when the sound wave strikes it. We will discuss this topic in more detail later.

Human Production and Detection of Sound

So how do humans produce sound? Clapping hands, stomping feet, popping knuckles, and belching are all different ways that sound is produced. When we talk or sing, we use the vocal cords, which are a set of membranes that vibrate when air rushes past them. The fundamental principle is that an object vibrates, pushing and pulling on the air around it, creating a pressure wave.

How do humans detect sound? The basic idea is that a pressure wave pushes and pulls on the eardrum, which is attached to other parts of the ear that eventually produce electrical signals that are sent to the brain.

Before we get into the physiology of speech and hearing, let's go over some other aspects of sound waves that are important for the human body.

Frequency Spectrum

Sound waves are produced when vibrating objects create pressure waves in air. However, not all sound waves can be detected by the ear. Only those sound waves with frequencies ranging from about 20 Hz to about 20,000 Hz can be heard. This range is called the audible region of the sound frequency spectrum. Sound waves whose frequencies are lower than 20 Hz lie in the infrasonic region and frequencies greater than 20,000 Hz are in the ultrasonic region. Figure 4.4 illustrates the sound spectrum. The human ear is most sensitive to frequencies in the range of 500 to 6000 Hz.

In the ultrasonic region, sound waves can be produced with frequencies ranging up to about a billion hertz, or one gigahertz (1 GHz). Dogs and cats can hear up to about 27,000 Hz, while dolphins and bats can hear up to about 100,000 Hz. Medical ultrasound equipment produces sound waves up to about 20 MHz. Ultrasonic frequencies tend to be absorbed easily so they do not travel far.

There is evidence that elephants and other large animals produce sound waves in the infrasonic region in order to communicate. Infrasonic frequencies also occur during earthquakes, volcanoes, and large weather events. Humans can feel infrasonic vibrations, which sometimes cause nausea and pressure on the body if they have large enough amplitude.

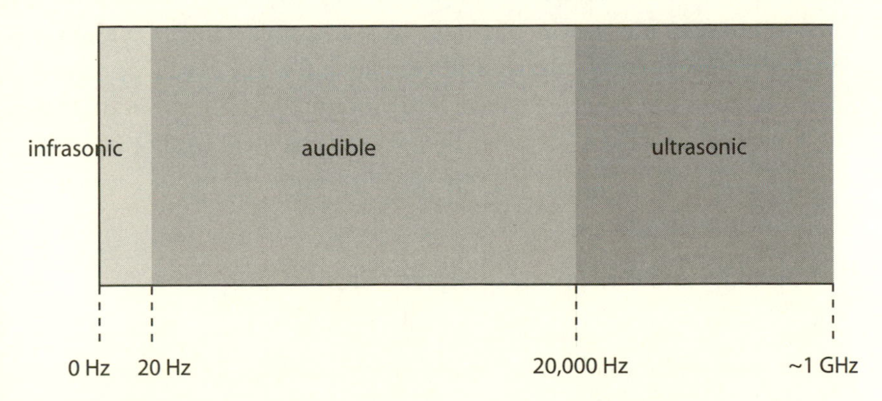

FIGURE 4.4. Sound frequency spectrum. It ranges from the infrasonic region through the audible region to the ultrasonic region. There is an upper limit in the ultrasonic region of about 1 GHz.

Infrasonic waves are not absorbed easily and can be transmitted through long distances. That is why when a car passes by on the street with speakers turned up you can hear the booming bass notes more easily than the higher notes. Elephants can communicate over distances of up to about five miles using infrasonic waves.

If the speed of sound is 343 m/s in air at 20°C, then the wavelength (with the corresponding frequency) of sound waves in the audible region ranges from about 17 m (at 20 Hz) to about 17 mm (at 20,000 Hz). By comparison, the size of the eardrum is about 10 mm, and the ear canal leading to the eardrum is about 25 to 26 mm long and about 6 to 8 mm in diameter.

Most sound frequencies produced by musical instruments fall within the range of a piano. The lowest note on the piano has a frequency of 27.5 Hz and the highest note is 4186 Hz. Middle C is 261.6 Hz and concert A is 440 Hz. By comparison, the typical male bass voice ranges from about 70 to 350 Hz, and the typical female soprano voice from about 220 to 1000 Hz. The violin ranges from about 200 to 2300 Hz, the guitar from about 80 to 700 Hz, the trumpet from about 160 to 1000 Hz, the flute from 260 to 2300 Hz, and the piccolo from 600 to 4200 Hz.

All of these instruments produce sound waves by causing an object to vibrate. In the human voice, it is the vocal cords that vibrate. In the piano, guitar, and violin, striking or plucking or bowing the string causes it to vibrate. In the trumpet and other lip reed instruments, the vibrating lip sets up a standing wave in the air inside the horn. In the flute, blowing across a hole causes a repeated pattern of swirling air into and out of the hole leading to vibration. In reed instruments, like the clarinet or saxophone, blowing across a reed causes it to vibrate. The frequencies of these instruments can be changed either by tightening a vocal cord or a string, or by changing the length of the tube as in a trumpet, flute, or clarinet.

Anatomy and Physiology of Speech Production

What is involved in the production of sound, or speech, from a physiological perspective? The main process involved in speech is the movement of air past the vocal cords. We will look briefly at how the air is forced to move and then more closely at the vocal cords.

The movement of air is caused by the organs that make up the respiratory system. The air is supplied by the lungs, which are filled and emptied when the diaphragm moves. This topic has already been discussed in chapter 2, but we will include some of that discussion here.

Recall that the diaphragm used in breathing is a large dome-shaped muscle attached to the bottom of the thoracic cavity that separates the cavity from the abdomen. When one inhales, the diaphragm moves downward and flattens out, causing an increase in the volume of the thoracic cavity, so that the pressure drops and air rushes into the lungs from outside the body. When one exhales, the diaphragm relaxes, moving upward into the thoracic cavity, causing the volume to decrease and the pressure to increase, forcing air to move out of the lungs into the atmosphere.

During normal breathing, inhalation and exhalation take about the same amount of time. On average, a person breathes about 12 to 15 times per minute so each cycle takes about 4 to 5 seconds. During exercise the respiratory rate may be as much as 30 to 40 times per minute. When a person is talking, the inhalation time is quite small, as short as half a second, so that the person can quickly catch a breath to continue speaking. During the physical act of speaking, the person is exhaling and the exhalation time can be as long as 5 to 10 seconds. If the person doesn't "catch a breath" then speech can actually go on for a much longer period of time. Trained vocal performers can sustain a note for up to a minute. By the way, a person can speak while inhaling, but it can be quite difficult and cannot be sustained for as long as when exhaling.

In order to produce speech, air from the lungs moves through the trachea, or windpipe, and then to the larynx, which contains the vocal cords or vocal folds (fig. 4.5). The larynx is also called the voice box and the Adam's apple. The reference is to Adam in the book of Genesis in the Bible where he has eaten of the forbidden fruit. Although the Bible does not specify the type of fruit, popular culture says that it was an apple. The name Adam's apple comes from the thought that the apple got stuck in Adam's throat, causing it to protrude from the front of his neck.

The larynx not only is involved in the production of speech, but also protects the trachea. In swallowing, the larynx moves upward behind the base of the tongue so that whatever is being swallowed goes down the esophagus to the stomach. If food does enter the trachea, a coughing reflex occurs: the larynx closes off the trachea and then suddenly opens

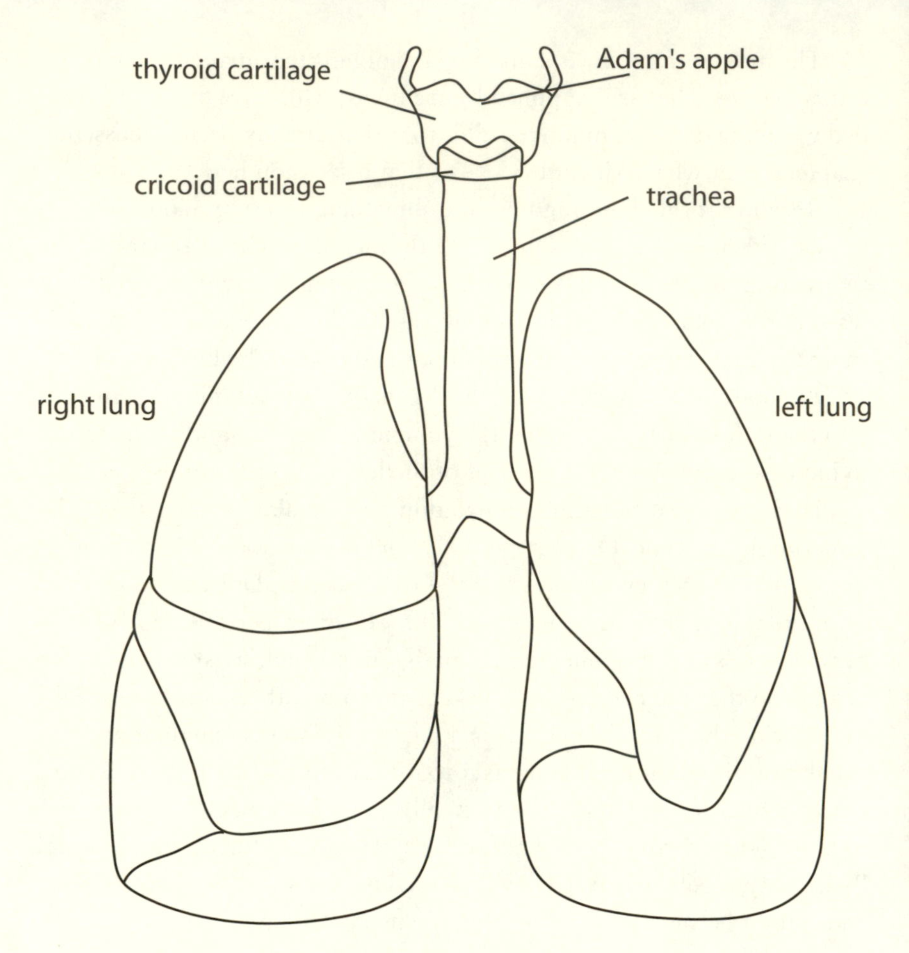

FIGURE 4.5. Anatomy of the larynx, trachea, and lungs.

so that a quick rush of air can force the food or other foreign matter out of the windpipe.

The larynx is a tube about 5 cm (2 in.) long composed of several cartilages, the largest of which are the arch-shaped thyroid cartilage and the ring-shaped cricoid cartilage (see fig. 4.5). The thyroid cartilage forms a shieldlike structure in the front and along the sides of the larynx for protection and support; the front of the thyroid cartilage protrudes from the throat, forming the Adam's apple. The cricoid cartilage provides structure at the base of the larynx and is attached to the trachea. Bones, muscles, and ligaments cause the cartilages and the vocal cords to move.

The vocal cords, also called vocal folds, are two flat, triangular muscles that project across the opening of the larynx (fig. 4.6). The folds are joined to each other at the front where they are connected to the inside of the thyroid cartilage. At the rear the two folds are connected to the arytenoid cartilages, which are attached to the cricoid cartilage. These cartilages allow the two folds to open and close and to change their tension in order to change the frequency of vibration of the folds. The folds and the opening between them are called the glottis.

The vocal cords, or glottis, can be open or closed under various circumstances. When the vocal cords are relaxed, the glottis is open. This occurs when a person is breathing normally or when whispering. The glottis is closed if there is straining, as when lifting a heavy object. When a person coughs, the glottis is closed and then suddenly opens, accompanied by a large rush of air. During speech, the glottis may be closed briefly before opening, as when sounding the letter "e" or "o," which is called a glottal stop.

Vibration of the vocal cords produces sound that can be varied according to frequency and loudness. Recall from equation (4), which we repeat here:

$$f = \frac{1}{2L}\sqrt{\frac{T}{\mu}}, \tag{6}$$

that the frequency of a vibrating string depends on the tension in the string and the mass density (or thickness) of the string. We can apply this equation to the vocal cords as well. First of all, the frequency, or pitch, of the

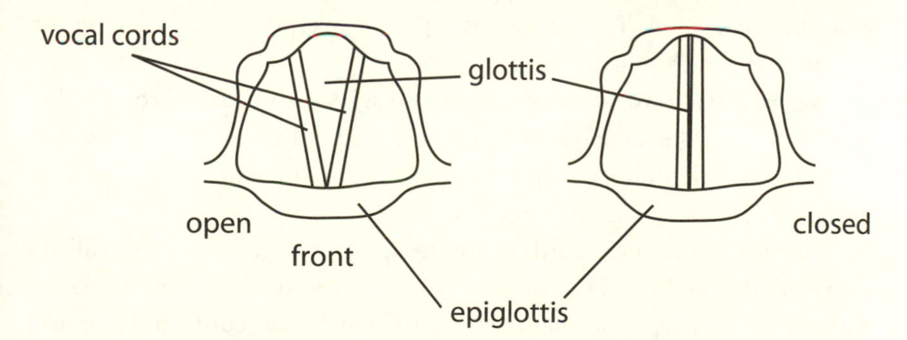

FIGURE 4.6. Sketch of the vocal cords and supporting structures showing the vocal cords open in the left figure and closed in the right figure.

voice depends on the tension in the cords. When the cords are tense, they vibrate faster and the pitch is higher. When they are more relaxed under less tension, they vibrate more slowly and the pitch is lower. Second, the frequency also depends on the size of the vocal cords: men typically have thicker vocal cords, which results in a lower voice; women's voices have a higher pitch because their vocal cords are thinner and vibrate faster.

There is another factor that contributes to the pitch of the voice; recall that larger objects tend to produce lower-frequency sounds, and smaller objects produce higher-frequency sounds. The frequency of the voice depends on the length of the vocal cords. Men typically have a large larynx with longer vocal cords; the result of this fact is a lower voice for men. Women's vocal cords are usually shorter because they have a small larynx, which causes their voices to have a higher pitch.

The loudness of the voice is related to the amplitude of vibration of the vocal folds. If there is a large rush of air from the lungs, the amplitude of vibration is large and the sound will be loud. When we are speaking softly, there is not a lot of air that rushes out of the lungs.

Why do the vocal cords vibrate when air flows past them? The flowing air helps to hold the cords open by pushing them upward away from the trachea toward the mouth. The tension in the cords helps to pull them back down, but they could stay in an equilibrium configuration without vibrating except for the *Bernoulli effect*. The Bernoulli effect says that there will be a drop in air pressure in the region where the air moves rapidly, so that there is a net pressure forcing the vocal cords downward toward the trachea. This action slows down the rushing of the air by a small amount, so the negative pressure goes away and the vocal cords move upward again. This process is repeated over and over, resulting in the vibration of the cords.

A simple demonstration of this effect occurs if you blow across a thin film of plastic, like a candy wrapper, or a blade of grass. You can make the film "squeal" quite loudly if you blow hard; you can change the frequency by changing the tension in the film.

Just above the vocal cords is another pair of membrane folds called the vestibular folds or the false vocal cords. These folds do not normally play a role in speech, but help protect the true vocal cords and help to close the glottis when swallowing.

When the vocal cords vibrate they produce sound, which usually consists of multiple frequencies that combine to form a complex periodic wave. The pharynx (located above the larynx), the mouth, and the nasal cavities all contribute to amplify and change the wave so as to impart a particular quality to the voice. The small differences in quality make each voice distinct, so that you can recognize whether your best friend is talking, or your mother is calling for you, or a voice synthesizer is simulating human speech. A website that has a wealth of information on voice and speech in general and about problems associated with voice and speech is www.nidcd.nih.gov/health/voice.[2]

Anatomy and Physiology of Hearing

What is involved in the detection of sound by the human body? What features of the human ear, from an anatomical and physiological perspective, are important for hearing? The primary feature involved in sound production is a vibrating object that pushes and pulls on the air around it, producing a pressure wave. So the reverse must be true for hearing: a pressure wave must strike an object (the eardrum), causing it to vibrate. These are the same features as in a loudspeaker and a microphone. Let's explore the ear in more detail.

The ear is composed of three parts: the external or outer ear, the middle ear or tympanic cavity, and the internal or inner ear (fig. 4.7). The outer ear is used to collect sound and channel it to the eardrum, the middle ear transmits the mechanical vibrations of the eardrum to the inner ear, and the inner ear converts these mechanical vibrations into electrical signals that go to the brain via the auditory nerve. The brain interprets these signals as sound.

The outer ear consists of two parts: the auricle, or pinna, and the auditory canal, or ear canal. The auricle is the visible part of the ear and is usually what we refer to when we mention the ear. The purpose of the auricle is to direct sound waves into the ear canal. It is also important in determining the direction of the source of sound. The ear canal is roughly cylindrical in shape and is about 25 to 26 mm (1 in.) long and 6 to 8 mm in diameter. It is open at the end where it is connected to the auricle, and is closed at the other end where it is connected to the eardrum. The ear canal is rather like a column of air that can resonate, similarly to the pipe

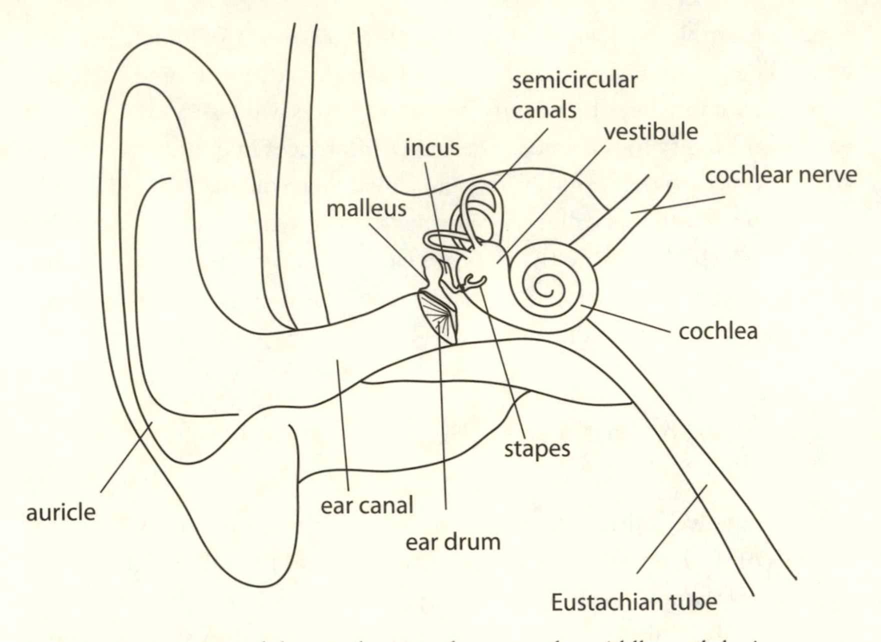

FIGURE 4.7. Anatomy of the ear showing the outer, the middle, and the inner ear.

of an organ, and it causes a magnification of the sound from the time it enters at the auricle until it strikes the eardrum. The eardrum also has glands that secrete ear wax that helps protect the ear by keeping out foreign matter such as bugs and dirt.

The middle ear consists of the eardrum and a set of three small bones called the *auditory ossicles*. It also contains the opening of the *Eustachian tube*, called the *pharyngotympanic tube* or *auditory tube*, because it connects the middle ear to the pharynx. Its purpose is to keep the middle ear at atmospheric pressure by opening briefly when a person yawns or swallows. This is the tube that we often hear popping when in an airplane or on a mountain road. The Eustachian tube also helps to drain mucus from the inner ear; it occasionally stops up, trapping bacteria in the middle ear, which causes it to become infected or inflamed (otitis media). Many children have ear infections; if the severity and frequency of these infections is a problem, then tubes can be surgically inserted into the eardrum to help equalize the pressure. In adults, if the ears become stopped up and won't drain, a doctor can puncture the eardrum with a needle and suck the mucus out.

The eardrum is also called the *tympanic membrane*. Remember we have talked about membranes or films that vibrate, either producing sound, like the vocal cords or a loudspeaker, or collecting sound as in a microphone. The eardrum is not flat, but has a concave surface to the outer ear and a convex surface to the inner ear, similar to a cone that has been nearly pressed flat.

The auditory ossicles are three small bones connected to each other. The first is called the *malleus*; it is shaped like a hammer whose "handle" is connected to the apex of the eardrum. Next is the *incus*, which is shaped like an anvil, and finally the *stapes*, which is shaped like a stirrup (such as on the saddle of a horse). The base of the stapes is aligned within the oval window that is the opening to the inner ear.

The purpose of the ossicles is to transmit the mechanical vibrations of the eardrum to the inner ear. The design of the ossicles creates pressure on the oval window (by the stapes) about 20 times larger than the pressure on the eardrum (connected to the maleus). This increased pressure is needed to cause wave motion in the fluid in the inner ear.

The inner ear consists of three main parts: the *cochlea*, the *vestibule*, and the *semicircular canals*. All three are actually cavities or channels in the temporal bone rather than stand-alone structures. The vestibule and the semicircular canals are used to help maintain balance of the body; they contain fluid that responds to movements of the head. The cochlea is a small chamber that is fluid filled and contains the cochlear nerve needed for hearing. Because the vestibule and semicircular canals are not used in hearing they will not be discussed in detail.

The cochlea is formed in the shape of a spiral. Its name comes from the Latin for "snail" because it is similar in shape to a snail's shell. The largest part of the spiral opens very close to the oval window and the apex of the spiral is the innermost part of this curved structure. The cochlea is divided into three separate chambers; the outer two chambers are connected at the apex and the center chamber, the cochlear duct, runs the entire length of the spiral toward the apex. The three chambers are separated by membranes, one of which, the *basilar membrane*, plays an important role in hearing.

When the stapes moves back and forth it produces a pressure wave in the fluid in one of the outer chambers; this pressure wave can move all the way to the apex, where it continues to move in the other outer chamber

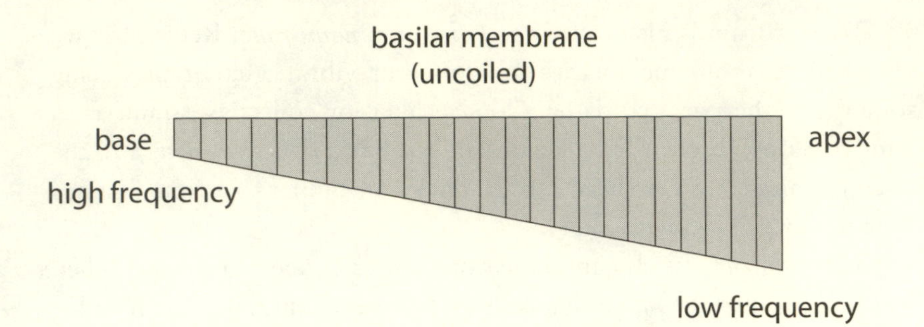

FIGURE 4.8. Sketch of the basilar membrane if it were to be stretched out. It consists of many fibers oriented perpendicular to its length that vibrate in response to stimuli from sound waves, and it is normally coiled inside the cochlea.

back to the base. At frequencies in the range of hearing (20 to 20,000 Hz) the pressure wave takes a shortcut from the first chamber through the cochlear duct to the other chamber. During this shortcut, the pressure wave causes the basilar membrane to vibrate.

The basilar membrane is composed of fibers of different lengths (fig. 4.8) that have different resonant frequencies. The fibers are short near the base of the cochlear duct and increase in length toward the apex. The short fibers respond best to high frequencies and the longer ones to lower frequencies. The vibrating fibers stimulate hair cells (stereocilia or microvilli) that are part of the *spiral organ of Corti*, which is connected to the auditory nerve. The electrical impulses produced by the bending of the cilia are sent to the brain to be interpreted as sound.

This discussion is a brief description of the hearing process. If you are interested in exploring this topic in more detail, check out a physiology book.[3]

Hearing Problems

Because of the many steps involved in hearing, a number of problems can arise that decrease the ability of an individual to hear properly. Remember that sound waves have to travel into the outer ear to the eardrum, where they are converted to mechanical vibrations that are amplified by the bones of the middle ear. Pressure is applied to the fluid in the inner ear, causing the cochlear membrane and then the hair cells to vibrate and stimulate the auditory nerve. Hearing loss is a decrease in the ability to receive and/or process sound and can occur anywhere along this path.

A buildup of ear wax is one of the first things to consider because it can block the ear canal. Some people produce excessive amounts of ear wax and they need to have their ears cleaned out often, perhaps several times a year. Tumors or abnormal bone growths can block sound as well. Swimmer's ear is an infection caused by water being trapped in the canal, resulting in swollen tissue so that sound does not properly move through the ear canal.

A perforated or ruptured eardrum can cause hearing loss because the eardrum is not able to vibrate properly. If a person is exposed to a very loud noise such as an explosion the eardrum can be damaged. A ruptured eardrum can also allow bacteria into the middle ear, which can cause an infection.

In the middle ear several problems may arise that cause hearing loss. Infection of the middle ear (otitis media) causes fluid to build up behind the eardrum so that it doesn't vibrate properly. It can also inhibit proper vibration of the ossicles. The middle ear can also become infected if the Eustachian tube becomes blocked, keeping mucus from draining properly. This can happen during a cold or if the adenoids, which sit in the upper part of the pharynx, become swollen. An abnormal growth, *otosclerosis*, in the middle ear can keep the ossicles from operating properly. Sudden changes in air pressure, as when changing altitude in an airplane, can cause air pressure in the middle ear to cause temporary discomfort and hearing loss.

Most hearing loss occurs in the cochlea when the pressure wave produced in the fluid of the inner ear does not stimulate the auditory nerves. Hair cells can be destroyed by a very loud noise or by long-term exposure to loud sounds, such as rock bands, lawnmowers, or airplanes. Also, as a person ages, there is a slow loss of hair cells. Medication and tumors can cause damage as can nerve degeneration and illnesses such as meningitis.

And then there is the brain. A loss of perception of sound can be caused by damage to the primary auditory cortex in the temporal lobe of the brain or to the auditory areas of the brainstem. A good source for discussion of hearing loss is the website www.mayoclinic.com/health/hearing-loss/DS00172.[4]

Intensity of Sound

All sound waves carry energy. You may have seen commercials where a woman sings a high note with such energy that it shatters a glass. The energy is transferred to the glass, setting up such a large vibration in it that it breaks. This transfer of energy is also apparent when the sound wave strikes the eardrum causing it to vibrate. Very loud sounds can cause the eardrum to rupture.

Rather than measure the amount of energy in a sound wave, it is more common to measure the *intensity* of a sound wave. Intensity I is defined as the amount of energy E in the sound waves that strikes a unit area A per time t. The energy per time is the power P. Mathematically, we can write the intensity as the following:

$$I = \frac{P}{A}. \qquad (7)$$

Typical units for intensity are watts per square meter (W/m^2) or watts per square centimeter (W/cm^2).

Sound intensity is related to the loudness of a sound wave. There are two values that serve as reference points: the threshold of hearing given as

$$I_0 = 10^{-12}\ W/m^2$$

and the threshold of pain, which has a value of

$$I_p = 1.0\ W/m^2.$$

The threshold of hearing is the intensity at which sound can just barely be heard. The threshold of pain is the point where the loudness of a sound begins to cause pain in the average person.[5] It is important to keep in mind that perception of sound is different for each person, so the intensity at which a person can hear or feel pain may be different. In addition, the intensity at which a person begins to hear or to feel pain depends on the frequency—sounds with frequencies at either end of the audible region generally need to have a higher intensity than the threshold of hearing in order to be heard. The values stated for I_0 and I_p are in wide use in the fields of acoustics and audiology.

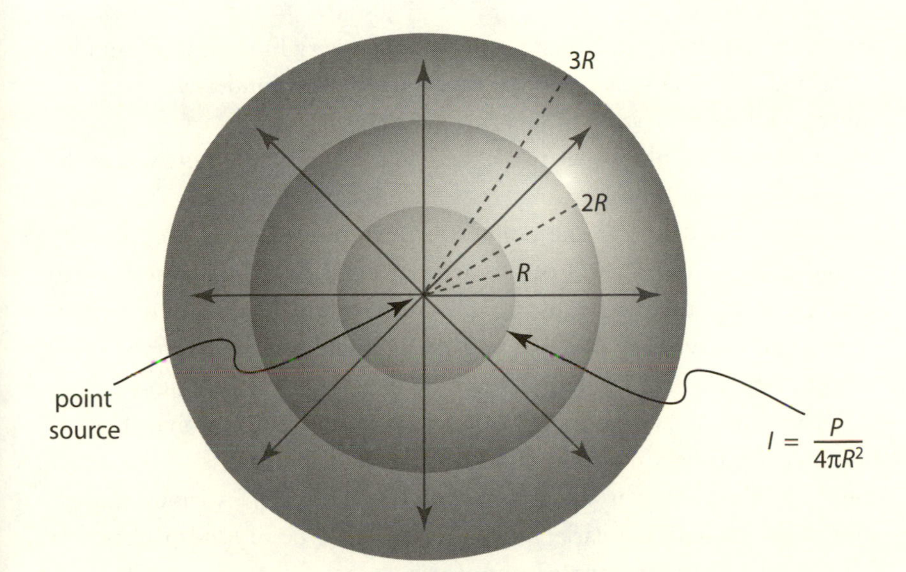

FIGURE 4.9. Sound radiates outward from a point source equally in all directions. If the point source has a power given by P, then at a distance R from the point source the power passes through an imaginary sphere whose area is given by $4\pi R^2$. The intensity I is equal to $P/(4\pi R^2)$, so the intensity decreases as $1/R^2$.

One method that can be used to determine the intensity of sound is to consider the object as a point source of sound. If a person claps hands or a drumstick bangs against a cymbal or a hammer strikes a nail, the source of sound can be considered as a point source. For a point source, the emitted energy spreads out equally in all directions (see fig. 4.9). The area into which the sound is passing is the surface of an imaginary sphere. So the intensity at a distance R from the point source can be calculated using the expression

$$I = \frac{P}{A} = \frac{P}{4\pi R^2}, \tag{8}$$

where $4\pi R^2$ is the surface area of a sphere of radius R.

From this expression we see that, as the distance of the observer from the source of sound increases, the intensity of the sound decreases. This makes sense from our experience. If we cannot hear someone clearly, we try to move closer to them. If sound from a loudspeaker at a rock concert is too loud, we may move toward the back of the theatre. If lightning strikes in your backyard, then you will hear a very loud clap of thunder, but if the lightning strikes several miles away, the thunder will be much quieter and perhaps will not be heard at all.

Equation (8) follows what is called an *inverse-square law*, where the intensity is inversely proportional to the square of the distance. This means that if the distance from the point source were to double, the intensity would decrease to only 1/4 of the original value. If the distance were to triple, then the intensity would be 1/9 of the original value. In the opposite case, if you were to move to half the original distance from the point source, then the intensity would increase by a factor of four times.

There are other sources of sound that are more directional. For example, cheerleaders use a megaphone that has a conical shape to direct their yells in particular directions. Loudspeakers have various designs that radiate sound in a variety of patterns. The type of loudspeaker that is used depends on the application: for a stereo system in a small room, for a large sound system in an auditorium or at an outdoor concert, or for a warning system in case of severe weather.

Sound Intensity Level

Suppose there are two sounds and the intensity of one is ten times the intensity of the other. The first one will be ten times louder than the second one, right? Actually, it turns out that the first sound is only about twice as loud as the second, as perceived by a person.

Because of these differences in perception, a different way to measure loudness is used. It is called *sound intensity level*. The intensity level is a logarithmic scale that converts the intensity to a scale more closely associated with loudness. The intensity level L_I is defined as

$$L_1 = 10 \log \frac{I}{I_0},$$

where once again I_0 is the threshold of hearing and has a value of 10^{-12} W/m^2. The unit for intensity level is the decibel, which is abbreviated as dB.

In this definition the intensity level is referenced to the threshold of hearing, because the threshold of hearing has an intensity level of 0 dB. Using the mathematical formula, if the intensity of a certain sound is equal to the threshold of hearing, that is if $I = I_0$, then $I/I_0 = 1$; the log of 1 is zero and ten times zero is still zero.

The range of hearing is from the threshold of hearing (10^{-12} W/m^2, 0 dB) up to and exceeding the threshold of pain. The threshold of pain is defined as 1 W/m^2, which corresponds to an intensity level of 120 dB. Normal conversation occurs at about 50 to 60 dB, with a soft whisper around 20 dB, and a rock concert at 110 to 120 dB. When people are exposed for long periods of time to loud sounds exceeding 90 dB, severe hearing loss can develop. Certain jobs require that workers be in situations of prolonged exposure to loud sounds: in a machine shop, or near airplanes, or when running lawn equipment. These workers are often required to wear hearing protection devices such as earplugs or noise-canceling headphones. We will look more closely at this topic in a moment.

An interesting feature of sound intensity and sound intensity level occurs when two sounds are compared: there is an additive factor of ten in intensity level for every multiplicative factor of ten in intensity. We can use the mathematical definition of sound intensity to see how this works by determining the difference between two sound intensities, $\Delta L_I = L_{I1} - L_{I2}$:

$$\Delta L_I = L_{I1} - L_{I2} = 10 \log \frac{I_1}{I_0} - 10 \log \frac{I_2}{I_0}$$

or

$$\Delta L_1 = 10 \log \frac{I_1}{I_2}. \quad (9)$$

Thus, if $I_1/I_2 = 10$ (one multiplicative factor of ten), then $\Delta L_I = 10$ dB (one additive factor of ten); or if $I_1/I_2 = 1000$ (three multiplicative factors of ten), then $\Delta L_I = 30$ dB (three additive factors of ten).

Conversely, for every increase in intensity level of 10 dB, there is a tenfold increase in intensity. To show this, let's use the formula for ΔL_I above and solve for I_1/I_2. We find

$$\frac{I_1}{I_2} = 10^{\Delta L_1/10}. \quad (10)$$

For example, an intensity level of 40 dB is ten times more intense than one of 30 dB. The difference in intensity level between a 92 dB source and a

72 dB source is 20 dB, so that the first is 100 times more intense than the second. The formula also works if there is a decrease in sound intensity level. For example, if $\Delta L_I = -30$ dB, then $I_1/I_2 = 1/1000$.

Earplugs

At most pharmacies, you can find earplugs. Earplugs make promises about reducing "noise pollution" or being "snore blockers." What they actually do is reduce the intensity of sound reaching the eardrum by absorbing sound waves as they try to enter the ear canal. Earplugs come in soft foam or moldable wax that can fit into the ear canal. Earplugs made of memory foam (which have a slow recovery after being squeezed) are the most effective as long as they are properly inserted in the canal. Earmuff-type hearing protectors fit over the ear and are even more effective.

An important feature of earplugs is the *noise reduction rating*, or NRR. The NRR rates the effectiveness of earplugs at decreasing the amount of sound that enters the ear. Typical values range from 12 to 33 dB. The NRR represents the amount of decrease in the intensity level of sound reaching the eardrum. In other words, an NRR of 29 dB means that $\Delta L_I = -29$ dB. Substituting this value into equation (10) gives $I_1/I_2 = 0.00126 = 1/794$, which means that the intensity of sound reaching the ear is reduced to about 1/800 times its original value. For earplugs with an NRR of 33 dB, $I_1/I_2 = 1/2000$.

The NRR is actually an average value obtained at specific frequencies of the audible spectrum. The U.S. government through its Occupational Safety and Health Administration (OSHA) in accordance with standards of the American National Standards Institute (ANSI) and the International Organization for Standardization (ISO) has established rules and regulations that prescribe how values of the NRR (and other rating systems) are determined.[6]

Earplugs are examples of passive methods used for noise reduction; other devices use active methods to cancel noise before it enters the ear canal. One method is based on an important property of waves that is seen when they overlap each other. If they are in phase with each other, so that a peak lines up with another peak, the result is a larger wave. However, if they are out of phase, when a peak of one wave lines up with a valley or trough of another, the two waves cancel each other. Noise-canceling headphones detect sound electronically and produce an out-of-phase sound wave; when the two waves are combined, they cancel each other so that the original noise no longer enters the ear canal.

Sound Pressure Level

Another scale used to compare sounds is called the *sound pressure level* or *SPL*. Recall that intensity is the rate of energy transfer per area in a sound wave and is measured in W/m^2. Sound intensity level compares the sound intensity of a particular sound to the threshold of hearing and is measured in dB. Both of these quantities use the word "intensity" and have to do with energy. But remember that a sound wave is also a pressure wave consisting of alternating regions of high pressure (condensations) and low pressure (rarefactions). The variation in pressure is actually small compared to the atmospheric pressure of the air in which the sound is moving. Atmospheric pressure is about 101,300 Pa (pascals). The deviation from this value at the threshold of hearing is only 2×10^{-15} Pa, whereas the deviation at the threshold of pain is about 20 Pa.

The sound pressure level compares the pressure deviation of a sound wave to the pressure deviation at the threshold of hearing. The mathematical formula is very similar to that of sound intensity level and is written

$$L_p = 10 \log \frac{p^2}{p_0^2} = 20 \log \frac{p}{p_0},$$

where $p_0 = 2 \times 10^{-5}$ Pa is the threshold of hearing. The unit is, once again, the decibel (dB). As it turns out, the sound intensity is proportional to the square of the sound pressure, so the sound power level of the threshold of hearing is 0 dB and of the threshold of pain is 120 dB just as it is for sound intensity level.

Noise

The term noise is usually meant to refer to unwanted sound or to sound that is loud and unpleasant. There are several variations of this term, one of which is *white noise*. White noise is produced when random frequencies of sound occur with roughly the same intensity averaged over time. The typical sound is that of a hiss similar to a waterfall (such as Niagara Falls) striking a large pool of water. The sound can be loud like a waterfall, or quiet, such as when an air conditioner or heater fan is blowing air through a room. This sound is often heard when you are tuning a radio and the receiver is between radio stations. White noise machines provide this sound to help cover other sounds so that a person can sleep or can focus on work.

Other sounds that are considered noise may be better defined in frequency, but often are annoying or unwanted. Lawnmowers, leaf blowers, car engines, police sirens, feedback in sound systems, and even many people talking at the same time produce very distinctive sounds with frequencies within a certain range that is often quite loud and unwanted. Earplugs or noise canceling headphones may need to be used to reduce the intensity of the sound entering the ear.

Hearing Aids

If a person experiences hearing loss, a hearing aid may be useful. Early hearing aids were called ear trumpets. They were carved from wood, or fashioned from seashells or from animal horns. They had a wide opening and a narrow base; the wide opening was pointed toward the source of sound and the narrow base placed to the ear. Sound was "funneled" to the ear by the ear trumpet. You may have used an improvised version of an ear trumpet—the funneling happens when you hold a cupped hand to your ear (the auricle) as you strain to hear well.

In addition to these early hearing devices, buildings and other structures were built in order to help people to hear well. Ornate furniture built with a reflecting dome or buildings with high dome-shaped ceilings help to reflect sound. I have toured numerous capitol buildings where the tour guide has pointed out the positions around a domed room where sounds, especially conversations, can be heard easily. Even long tubes were placed strategically in buildings so that someone could listen in on the conversation of another party. Electronic microphones began to emerge in the late 1800s and early 1900s. In 1933 a device was invented that could transmit sound waves through the skull. Bone conduction of sound is still in use today.

Modern hearing aids are devices that amplify sound by electronic means. A hearing aid collects sound using a microphone (or several microphones) and converts it to an electronic signal, which is used to drive a speaker that produces sound with a larger intensity. Some hearing aids amplify the sound with almost exactly the same frequencies as the original. Others modify (modulate) the sound depending on how it is made and the needs of the person by changing the frequency of the sound, or by filtering out unwanted noise.

Cochlear implants are different from typical hearing aids. A person who needs a cochlear implant has severely limited hearing and could be completely deaf. Most hearing aids amplify sound for the person to hear, but cochlear implants do not. Rather, they provide electrical stimuli directly to the auditory nerve, which the brain interprets as sound. A cochlear implant consists of a microphone, a speech processor, and a transmitter that are attached behind the ear outside the head, a receiver just beneath the skin behind the ear, and an array of electrodes that are surgically inserted into the cochlea. A great deal of therapy may be required to understand the new "sounds" as interpreted by the brain. Two websites that can provide much more information are www.nidcd.nih.gov/health/hearing[7] and www.asha.org/public.[8]

Ultrasound Imaging

An ultrasound image is what most parents get to see when they are shown a "picture" of their unborn baby for the first time. Because of the nature of sound waves, ultrasound imaging has become an important tool for the medical professional, particularly in the area of obstetrics. X-rays can cause damage to healthy tissue and MRIs require that patients be subjected to very large magnetic fields, so ultrasound technology has been developed as a safe method for peering into the body. The images are not as good as CT scans (x-rays) and MRIs, but again the ultrasound method is much safer.

To produce an image the operator uses a handheld device that emits ultrasound waves; typical frequencies are within the range of 1 to 18 MHz (megahertz). The waves penetrate into the body and encounter the boundary between various body tissues. Some of the waves are reflected back to the handheld probe while others continue moving through the tissue or are scattered to other parts where again some are reflected back to the probe. Ultrasound waves do not penetrate bones because they are easily absorbed, but work rather well for looking at soft tissue.

Based on the speed of sound in human tissue, which is about 1500 m/s, the ultrasound machine calculates the penetration depth of the waves and then forms an image of the tissue. The handheld device can be placed in a variety of locations and orientations so that different sections of the body can be imaged from different angles. The images are displayed in real time on a video monitor.

Because of the real time nature of the images, they can show movement of the internal organs and even blood flow. Doppler ultrasound is a newer technique that can measure speed and direction of blood flow in arteries to determine blockages due to blood clots or narrowing due to buildup of plaque.

The resolution of images can be as small as a fraction of a millimeter. As an example suppose that the frequency of the sound wave is 3 MHz (3×10^6 Hz) with a speed of 1500 m/s. Then, using the expression $v = f\lambda$, we find that the wavelength $\lambda = v/f = (1500 \text{ m/s})/3 \times 10^6 \text{ Hz} = 0.0005$ m = 0.5 mm. With a wavelength of 0.5 mm, the ultrasound will provide details to that size or slightly smaller. For more detail, higher-frequency machines can be used because the waves have shorter wavelength. There is a trade-off, however, because higher -frequency waves do not penetrate as far as lower-frequency waves.

There are medical applications other than imaging that involve ultrasound waves. Absorption of ultrasound can result in localized heating of tissue to aid in cell repair or to treat pain. Ultrasound can be used in surgical applications such as the ablation (or destruction) of cancerous tissue. It can also be used in lithotripsy, which is a method to break up kidney stones or gall stones.

Summary

The ability to hear is an important sense of the human body. The production and detection of sound are important tools for communication. However, even for individuals who cannot speak or hear, there are other ways to communicate using sign language or technology. The unique properties of sound waves make them useful, even at frequencies outside the range of normal human hearing. There is much more to this topic than I have presented here. Experts study extensively to learn about speech and hearing so that they can become audiologists, speech therapists, surgeons, and so on to help people with hearing or speech problems. Others take a keen interest in the production and detection of sound through electronic applications. And still others are thrilled to be able to appreciate music by playing an instrument or singing, or simply enjoying a good concert.

FIVE

Electrical Properties and Cell Potential

Have you ever scuffed your feet across the carpet in your house and then reached for the door knob and ended up getting zapped? Almost every time I get out of my car, I touch the door to close it and I get quite a shock. The usual explanation for these phenomena is "static electricity." There is something about electricity that makes it scary, yet fascinating, as we observe the many ways it is used in our lives. Electrical devices such as computers and televisions come with all sorts of warnings about electrical hazards. And yet, electrical signals are running through our bodies all the time, allowing us to sense heat, cold, pain, thirst, hunger, taste, sound, and light, and also to respond to those sensations.

In this chapter we will look at some basic physics principles in the area of electricity and how these principles apply to the human body. Almost all operations of the body, such as motion, sensation, and regulation, involve electrical signals in the body. We will discuss the nervous system and examine the motion of charge in neurons. We will also look briefly at measurements of electrical activities of the body that are used for medical diagnosis of the brain, the heart, and muscular activity.

Electric Charge

Matter is made up of atoms, which are composed of electrons, protons, and neutrons. Electrons and protons are charged and neutrons have no charge. Charge is considered a fundamental property of an object. (Mass is another fundamental property that is related to the amount of matter in an object.) The charge of a proton has the specific value of 1.6×10^{-19} coulomb (C); this value is called the fundamental unit of charge. The coulomb is the SI unit of charge. The symbol e is often used to represent this value. Electrons have the same value of charge as a proton, but their charge is negative. So, with $e = 1.6 \times 10^{-19}$, the charge of a proton is $+e$, the charge of an electron is $-e$, and the charge of a neutron is zero.

If there is a collection of two or more electrons and/or protons, the overall charge of the collection can be determined by adding the individual charges. For example, if there are ten protons in a collection then the charge of that collection is $+10e$. If there is a collection of five protons and five electrons then the total charge is zero, and the collection is considered electrically neutral. The symbols q and Q will be used to represent a general value of charge whether it refers to a single charge or a collection of charges.

Atoms are typically neutral, which means there are equal numbers of protons and electrons; the number of neutrons plays no role in determining the charge of the atom. A charged atom is called an ion: this charge is a result of an imbalance of protons and electrons. If the ion has an overall positive charge, then it is a *cation*; if it has an overall negative charge, then it is an *anion*. For example, the most common form of potassium, whose chemical symbol is K, has 19 protons, 19 electrons, and 20 neutrons, and is electrically neutral. If one of the electrons is removed, the potassium cation K^+ is produced, which has a charge of $+e$. Another example is chlorine (Cl), which in its neutral state has 17 protons, 17 electrons, and 18 neutrons. In many reactions, chlorine will take on another electron, becoming the chlorine anion Cl^-.

In most situations, excess charge in an object arises by addition or removal of electrons. This is because electrons are not as tightly bound in the atom as protons. Protons are held very tightly together in the nucleus and require a lot of energy to separate. Electrons orbit the nucleus with energies that are much smaller and, thus, they are more easily removed. For metals, there is a huge number of electrons that can move—they can be easily removed, or more electrons can be easily added. In fact, physicists sometimes call this a "sea" of electrons. When insulating objects are rubbed together, for instance when you scuff your feet across the carpet, electrons may be removed from one object and added to another object. There are some special situations that arise, particularly in high-energy physics experiments, where isolated protons are moved around, but these are somewhat rare events.

Objects can also become charged when ions are added or removed (remember, however, that ions themselves are created by the addition or removal of electrons). In the human body, for example, cations (positively charged) may be in one area giving it an overall positive charge, and anions (negatively charged) in another giving it an overall negative charge. Ionic solutions in the body are often electrically neutral with equal concentrations of positive and negative ions.

The total charge q of an object or collection of charges must be an integer (N) multiple of e; thus $q = \pm Ne$, or $\pm e$, $\pm 2e$, $\pm 3e$, $\pm 4e$, and so on. Another way of saying this is that charge is *quantized*, which means that it is not continuous, but occurs in discrete, separate values. There are some situations where charge is either $\pm \frac{1}{3}e$ or $\pm \frac{2}{3}e$, but these values apply to particles called quarks. It doesn't make sense to discuss a net charge of $q = 1.0 \times 10^{-18}$ C because $q/e = 6.25$ is not an integer multiple of e. However, it does make sense to discuss a net charge of 1.0×10^{-9} C because $q/e = 6.25 \times 10^{9}$, which is an integer (a very large integer!).

We will sometimes talk about the amount of charge that moves across the cell membrane of a neuron (nerve cell) in the body. To find the total charge, we use the expression $q = \pm Ne$. For example, suppose two million (2×10^{6}) sodium ions (Na^+), each with a charge of $+e$, move from one side of the cell membrane to the other. Then the total charge that moves is

$$q = +Ne = +2 \times 10^{6}e = 2 \times 10^{6}(1.6 \times 10^{-19} \text{ C}) = 3.2 \times 10^{-13} \text{ C}.$$

This changing amount of charge in a cell produces an electrical signal that may be detected by the brain to produce a certain action of the body.

The total charge of an isolated system remains constant. This concept is called the *conservation of charge*. For example, suppose you take an air-filled balloon and rub it on your hair. Electrons are transferred from your hair to the balloon. The balloon gains a net negative charge and your hair ends up with a net positive charge, but the total charge of the system (hair and balloon) remains constant. Charge is conserved in all situations: in small chemical reactions, or in large thunderstorms with lots of lightning, or during the formation of stars.

Electric Force

How do we know that there are positive and negative charges? The answer comes from the concept of forces between charges. Charges exert forces on each other, and the behavior of these charges is different for different types of charge. After rubbing a balloon in your hair, you can get it to stick to a wall because of the transfer of charge. Or you can hold the charged balloon up to your hair and feel your hair standing up as it is attracted to the balloon. Perhaps you have taken the laundry out of the dryer and had to peel a sock off a shirt—they are stuck together because they become charged when they tumble in the dryer.

Recall from chapter 1 that when a force occurs, there must be two objects involved and these two objects exert forces on each other. Newton's third law says that if the first object exerts a force on the second object (action force), then the second object exerts a force (reaction force) on the first that is equal in magnitude, but opposite in direction.

Electric forces involve pairs of charges, both positive, both negative, or one of each (fig. 5.1). These forces occur for individual charges such as protons and electrons, for collections of charges such as positive and negative ions, and for charged objects such as balloons and clothing. Charges that have the same sign (positive or negative) repel each other, whereas charges with opposite signs attract each other ("opposites attract"). If two isolated positive charges are placed near each other and released, they fly apart. The same is true for isolated negative charges. However, if a positive charge is placed near a negative charge and they are released, they move toward each other because they are attracted to each other.

The size or magnitude of the force F between two charges depends on the size of the charges, q_1 and q_2, and the distance d separating the charges. The force is modeled using the mathematical expression

$$F = k\frac{|q_1 q_2|}{d^2}, \qquad (1)$$

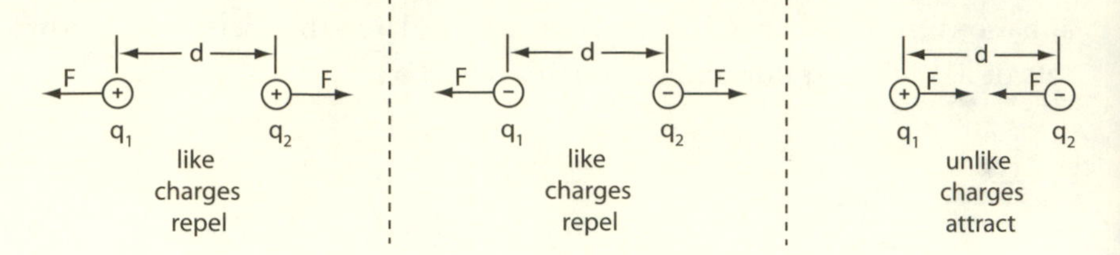

FIGURE 5.1. Properties of electric force. The electric force between a pair of charges, q_1 and q_2, obeys the law of charges, which states that like charges repel and unlike charges attract. The size of the force is determined using Coulomb's law, which depends on the distance *d* between the charges.

where *k* is called the electrical constant and is equal to 9.0×10^9 N m^2/C^2. The absolute value symbol is used to indicate that the signs of the two charges are not important in determining the size or magnitude of the force. (Force is a vector quantity, and often in physics the direction of a vector is expressed using a positive or a negative sign. In this discussion, I want to discuss the size of the force and explicitly state the direction when it is needed.) Equation (1) is called Coulomb's law and the electric force between charges is called a Coulomb force.

The expression shown above is the force between any two charges. But keep in mind that, for a collection of charges, there are forces between all the pairs of charges. Each charge has several forces acting on it, and it is the combination of all of these forces that will cause the charge to move or not. In an atom, the protons and electrons that make up the atom all exert forces on each other. In neutral atoms, these forces act in a way that allows us to consider the atom as a unit with no net charge. For ions, the forces allow us to consider the entire atom as a single charge of $\pm e$, $\pm 2e$, and so on. For larger objects with millions or billions of excess charges, the model applies if we are sufficiently far away from the object.

Let's look at an example related to the human body. Carbon is a common atom in the body, with the nucleus of carbon containing six protons and six neutrons. On average there are six electrons associated with each carbon atom, either orbiting the nucleus or shared with other atoms in chemical bonds. With all of these charges (protons and electrons) there are many electrical forces acting. However, we'll focus on the forces between the nuclei of two carbon atoms. Each nucleus has a charge of +6*e*, and we assume that they are separated by 0.5 nm. The force that each nucleus exerts on the other is

$$F = k\frac{|q_1 q_2|}{d^2}$$

$$= (9.0 \times 10^9 \ \text{N m}^2/\text{C}^2)\frac{[6(1.6 \times 10^{-19}\,\text{C})][6(1.6 \times 10^{-19}\,\text{C})]}{(0.5 \times 10^{-9}\,\text{m})^2}$$

$$= 3.3 \times 10^{-8}\ \text{N}.$$

This force looks quite small, particularly when compared to forces on larger objects (a few newtons to thousands of newtons). However, if the nucleus of a carbon atom were able to move under this amount of force, the acceleration would be huge because the mass of the carbon nucleus is very small, about 2×10^{-26} kg (recall Newton's second law: $a = F/m$).

Suppose there is a region in the human body where there are a large number of positive ions and another region where there are many negative ions. If a positive ion is placed in the space between these two regions, it will tend to want to move away from the positive region because of the repulsive force acting on it. In addition, it will be pulled toward the negative region because it is attracted to it (fig. 5.2). For the human body, ions such as sodium (Na^+), potassium (K^+), and chlorine (Cl^-) are the main charges involved. They tend to separate in different regions, such as inside and outside a cell, separated by a membrane. Electric forces act on these ions causing them to move into and out of the cell through the membrane. There are other types of ions present in and around the cell with electrical forces acting on them as well. But electrical forces are not the only causes of ion motion into and out of the cell. We will return to this topic in more detail later.

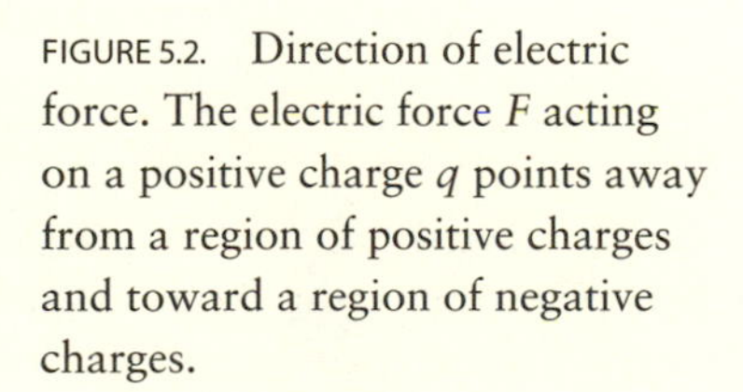

FIGURE 5.2. Direction of electric force. The electric force *F* acting on a positive charge *q* points away from a region of positive charges and toward a region of negative charges.

Electric Field

The electric force that acts on a charge is caused by another charge or a collection of charges that may be outside the field of view. Now suppose that the original charge is removed. The other charge or charges are still present, so when the charge is removed, there must be something that exists at that point in space: the *electric field*.

An electric field is produced by a charge (or collection of charges) in the region around the charge (or charges). There is a direction associated with this electric field; it is the same as the direction of the force that would act on a positive charge, called a test charge, placed in the region around the first charge (fig. 5.3). A positive test charge is repelled by a positive charge and attracted to a negative charge, so the electric field points away from a positive charge and toward a negative charge.

This concept is helpful when we consider what happens to charges in the human body. If a collection of positive charges exists on the outside of the cell membrane and a collection of negative charges exists on the inside, then an electric field exists within the membrane itself and points from the outside toward the inside. If a positive sodium ion (Na^{+}) sits in

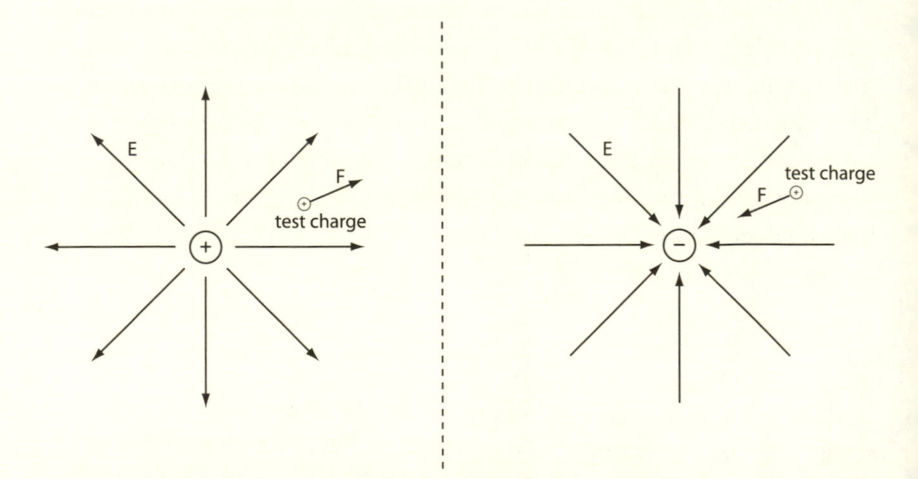

FIGURE 5.3. Electric field and electric force. The electric field *E* near a charge points away from a positive charge and toward a negative charge. The force *F* on a positive test charge points in the same direction as the electric field.

the membrane, there is an electric force causing it to want to move toward the inside of the cell. The existence of electric fields and forces in the neuron is a key point in understanding how nerve cells function.

The relationship between force and field is easy to express mathematically. It is

$$F = |q|E. \tag{2}$$

This expression says that if we place a charge q at a point where the electric field is E, there will be a force acting on that charge given by F. Conversely, if you know the force F that acts on a charge q, the electric field at the point where the charge is located is given by E. The interpretation of this expression depends on what is known about a situation and what needs to be calculated. The absolute value of q means that we don't have to worry about the sign when determining the size of the force.

The electric field for a single charge q can now be stated in mathematical terms. It is

$$E = k\frac{|q|}{d^2}. \tag{3}$$

This expression says that the electric field at a distance d from the charge q is directly proportional to the charge—the larger the charge, the larger the field. The direction of the field, as stated previously, is away from the charge if it is positive and toward the charge if it is negative.

Note that equation (3) is similar to equation (1) for the force between two charges. That is because the field created according to equation (3) causes a force to act on another charge. For example, suppose equation (3) is written for a charge q_1 and substituted into equation (2), but let equation (2) be the force exerted on a second charge q_2. The result is equation (1), which is the force that q_1 exerts on q_2. However, keep in mind Newton's third law—the charge q_2 exerts a force of the same size on q_1. Equation (3) can also be used to determine the electric field caused by q_2 and substituted into equation (2) to determine the force on q_1.

Now, all this may be a bit confusing and you may think that too much math has been thrown at you. However, there are two main points to get out of this: (1) charges produce electric fields and (2) a force acts on a charge that is placed in an electric field. The first point is true no matter the configuration of the charge, whether it is a single electron or proton, an ion, or charge distributed throughout a larger region, such as a cell or a balloon. The second point helps us to determine whether a charge will move or not. These points will be helpful when we talk about electrical properties of the body.

Electric Potential Energy

Potential energy is the energy of position or configuration. An object has potential energy because of its interactions with other objects around it. We spent a good bit of time in chapter 3 talking about gravitational potential energy, which is the energy of an object based on the gravitational force that the earth exerts on the object. In this chapter we want to describe briefly *electric potential energy*. This is the energy of a charge (or collection of charges) as it interacts with other charges because of the electric forces that are exerted on it. Recall that energy was defined as the *capacity to do work*. Potential energy has the potential to do work, and can be converted into other forms of energy that can do work. Charges that have electric potential energy may be able to move in order to do work or accomplish a particular task.

Suppose you have a positive charge located at a certain point and you want to bring another positive charge near it. Because the two charges repel each other, you have to push on the second charge to bring it near the first charge: you have to do work on the charge. The charge configuration gains potential energy as you bring the charges near each other. If the second charge is suddenly released, it will move away from the first charge. The potential energy of the original configuration is converted into kinetic energy of the moving charge—the potential energy decreases while the kinetic energy increases.

For the two charges just described we can write a mathematical formula that quantifies the potential energy:

$$U = k\frac{q_1 q_2}{d}. \tag{4}$$

You may notice that this expression is similar to the force expression in equation (1) (Coulomb's law). However, the electric potential expression [eq. (4)] does not have absolute value symbols, which means that the potential energy can be positive or negative depending on the sign of each charge. More specifically, if both charges are positive or both negative, the potential energy is positive; this indicates that it takes work to bring these two charges together because of the repulsive forces. If one charge is positive and the other negative, the potential energy is negative; they are attracted to each other and you have to apply a force to keep them apart. Recall that the absolute value signs in equation (1) are needed to determine the size or magnitude of the force.

The other difference between the two expressions is that the Coulomb force is inversely proportional to the square of the distance separating the charges, or $F \propto 1/d^2$, whereas the potential energy is inversely proportional to distance to the first power, or $U \propto 1/d$. In both cases, however, as the distance between the charges increases, the force and the potential energy get smaller and smaller (the Coulomb force decreases much more rapidly). Notice that the force and the energy both approach zero when the two charges are very far apart, because then they have very little interaction. Theoretically, the charges have to be infinitely far apart to exert no force on each other and to have no potential energy. This idea is true no matter whether the charges are both positive or both negative, or if they have opposite signs.

Charges move when placed in electric fields because of electric forces that act on them. They prefer to move on their own in the direction of *decreasing* potential energy. Work has to be done on them to move them in the direction of *increasing* potential energy. An example using gravitational potential energy occurs when you lift an object from the ground to a certain height and then drop the object. You do work on it to lift it, and when released it falls back to the ground. This example is similar to moving unlike (one positive and one negative) charges apart. You have to do work on them to separate them, but if they are released they will move toward each other.

Another example referring to objects on the earth considers a helium-filled balloon. This balloon prefers to move upward away from the earth, and it takes work to move it downward toward the earth. This example is similar to bringing two positive charges together or two negative charges—you have to do work on them to bring them together, but when they are released they will move away from each other.

In the cell membrane of the neuron, if a positive ion were free to move, it would want to move from the outside toward the inside of the cell. This is because the positive ion has more potential energy when it is outside the cell near other positive ions. It has lower energy inside the cell where it is near negative ions.

Electric Potential

Another quantity that is important in the study of electrical properties is called the *electric potential*. You may already be familiar with this quantity, if not the term. Another word that is often used for electric potential is voltage, such as the voltage of a battery. A car battery has a voltage of 12 V; a flashlight battery has a voltage of 1.5 V (fig. 5.4).

Electric potential is not the same as electric potential energy. Recall that the potential energy of a charge results from its interaction with another charge or a collection of charges. If the first charge is removed, then the other charge or charges produce an entity that remains at the point where the original charge was located. That entity is the electric potential.

As it turns out, the energy of the original charge depends on its size: the larger the charge, the greater the energy. The electric potential removes the dependence of the size of the charge from the picture; the electric potential V is defined as the energy U of a charge q divided by the charge, or

$$V = \frac{U}{q}. \tag{5}$$

From this expression we see that the SI unit of *volt* for electric potential is identical to a joule per coulomb. Thus 1 V = 1 J/C. Note that the symbol V for electric potential is the same as the symbol for the unit of volt. Sometimes in other literature you may see the symbol for electric potential written as the letter E. This E stands for "EMF" or electromotive force, which is not really a force, but an electric potential.

Electric potential is caused by a charge or a collection of charges just as electric field is caused by a charge or a collection of charges. Mathematically, the expression for the electric potential is written

$$V = k\frac{q}{d}. \tag{6}$$

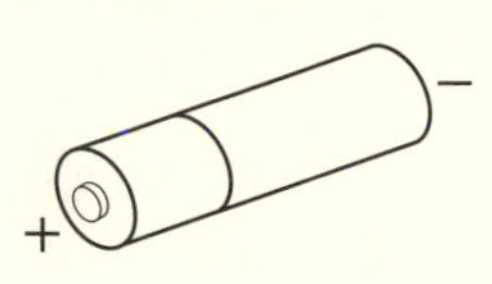

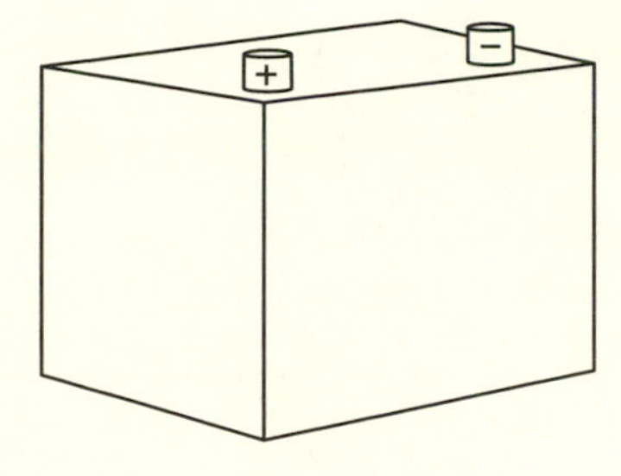

FIGURE 5.4. The electric potential difference, or voltage, is measured between the positive and negative terminals. For a flashlight battery, such as an AA, AAA, or D battery, the voltage is 1.5 V, while for a car battery it is 12 V.

This expression indicates that near a positive charge the electric potential is positive, and near a negative charge the potential is negative. The potential is zero at very large distances from the charge (fig. 5.5). When there is a collection of positive ions outside a cell and negative ions in the cell, the outside has a positive electric potential compared to the inside, which has a negative potential.

The previous statement brings us to an important concept: the electric potential is a quantity that has an arbitrary reference point; the location where the electric potential is equal to zero can be arbitrarily chosen. In equation (6) the zero value for electric potential is *chosen* to be where d is infinitely large, that is, at positions that are very far away from the charge. Because its actual value is arbitrary, often we are more interested in *changes* in electric potential between two locations than its value at either location. The term used to describe this concept is *potential difference*.

This idea of potential difference occurs in many applications. You are probably very familiar with 1.5 V batteries such as AA or AAA batteries. These batteries have a positive end (or terminal) and a negative end. The potential difference between the two terminals is 1.5 V: the positive terminal is 1.5 V higher in electric potential than the negative terminal, or the negative terminal is 1.5 V lower than the positive terminal. In many applications, the negative terminal is considered 0 V (sometimes called

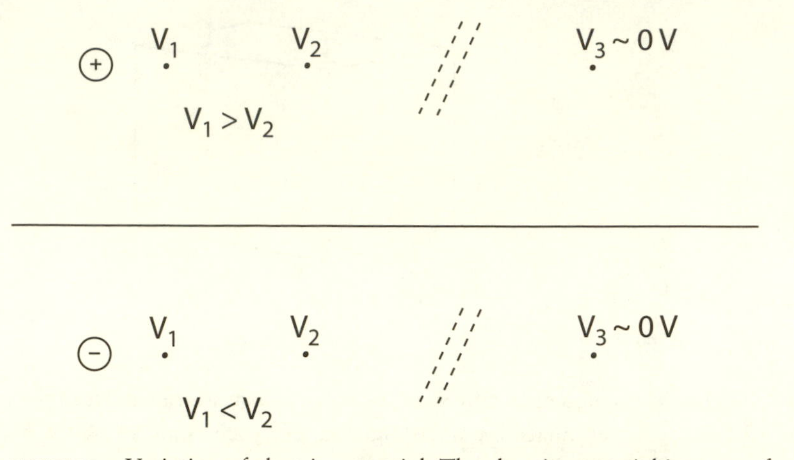

FIGURE 5.5. Variation of electric potential. The electric potential increases for locations closer to a positive charge and decreases for locations closer to a negative charge. In the figure, the potential V_1 is greater than V_2 near the positive charge, but V_1 is less than V_2 near the negative charge. The electric potential given by V_3 approaches zero for locations farther and farther from a positive or a negative charge.

"ground"), so that the positive terminal is at +1.5 V. Sometimes it may be that the positive terminal is connected to "ground" at 0 V, which makes the electric potential of the negative terminal equal to −1.5 V.

For the neuron, the outside of the cell has a higher potential than the inside. The potential difference across the membrane, called the *membrane potential*, is measured with the outside of the cell taken to be the reference point, or ground. We will discuss the membrane potential in much more detail later.

Electric Potential Difference and Potential Energy

Suppose a charge q were to move (or be moved) between two locations that have a potential difference ΔV, say across a cell membrane. A variation on equation (5) allows us to determine the change in the potential energy of the charge by using the expression

$$\Delta U = q\Delta V.$$

The charge gains or loses energy depending on the sign of the charge and the sign of ΔV.

Recall that objects tend to want to move on their own from regions of high to regions of low potential energy. If the charge q is positive, then it will want to move in the direction of decreasing electric potential, away from positive charges and toward negative charges. However, if q is negative, it will want to move in the direction of increasing potential, toward positive charges and away from negative charges. For the cell membrane, positive ions prefer to move from the outside of the cell surface where there is net positive charge with a relatively high electric potential, toward the inside of the cell where there are few positive charges and a lower potential. Note that this idea of movement to regions of lower potential energy is consistent with the direction of the forces acting on these charges.

What happens if an object is moved so that its potential energy increases? Work must be done on the object by another force. For example to lift a bowling ball, we must apply an upward force on it. When we move it from the ground to our waist we do positive work on it and the ball gains gravitational potential energy. The same idea applies to charges that gain more electric potential energy when moved: there must be other forces that cause them to move. In the cell membrane, positive ions tend to want to move from outside to inside (because the inside has a lower electric potential), losing electric potential energy as they move. However, in the normal resting state, the membrane potential is set up so that when positive ions are forced to move from inside the cell to outside (by what is called the *sodium/potassium pump*), they gain potential energy. When the neuron is operating properly, ions move back and forth across the membrane, gaining or losing energy, and producing a membrane potential that changes over time.

Capacitance

Suppose there are positive charges in one region and negative charges in another. We can talk about the electric potential and the electric potential energy of this configuration, as well as the electric field produced by the charges and the forces that are exerted on the charges by each other. There tends to be a limit on the amount of charge that can be stored in such a configuration because of the forces involved. To accomplish this separation of charge a device called a *capacitor* is often used.

A typical capacitor used in many electronic applications consists of two metal plates that face each other and are separated by a small distance. The capacitor is charged when the plates are attached to the terminals of a battery: one plate to the positive and the other to the negative terminal. Electrons move from one plate to the positive terminal of the battery leaving that plate with a net positive charge; electrons are added to the other plate from the negative terminal of the battery giving that plate a net negative charge. The total charge of the capacitor is actually zero, with an equal amount of positive charge on one plate and negative charge on the other.

The amount of charge transferred to the capacitor depends on a property called the *capacitance*. The capacitance of a capacitor is a measure of its ability to store charge: the larger the capacitance, the larger the charge that can be stored on it. The capacitance depends on physical properties of the capacitor, such as the area of the plates, the separation of the plates, and any material between the plates. An expression that helps us to model the dependence on these properties is

$$C = \varepsilon \frac{A}{d},$$

where C is the capacitance, A is the area of the plates, d is the separation, and ε is called the dielectric permittivity of the material between the plates (fig. 5.6). Even if there is no material between the plates, they can still hold charge; the dielectric permittivity is then called the permittivity of free space and has a value $\varepsilon_0 = 8.85 \times 10^{-12}\ C^2/(N\ m^2)$. The relationship between ε and ε_0 is given by the expression

$$\varepsilon = \kappa \varepsilon_0, \tag{7}$$

where κ is called the dielectric constant. The dielectric constant of vacuum is equal to 1, whereas for human tissue it is in the range of 30 to 80. Note that for larger values of the dielectric constant the capacitor has a larger capacitance.

The amount of charge Q that the capacitor can hold ($+Q$ on one plate and $-Q$ on the other) depends not only on the capacitance C, but also on

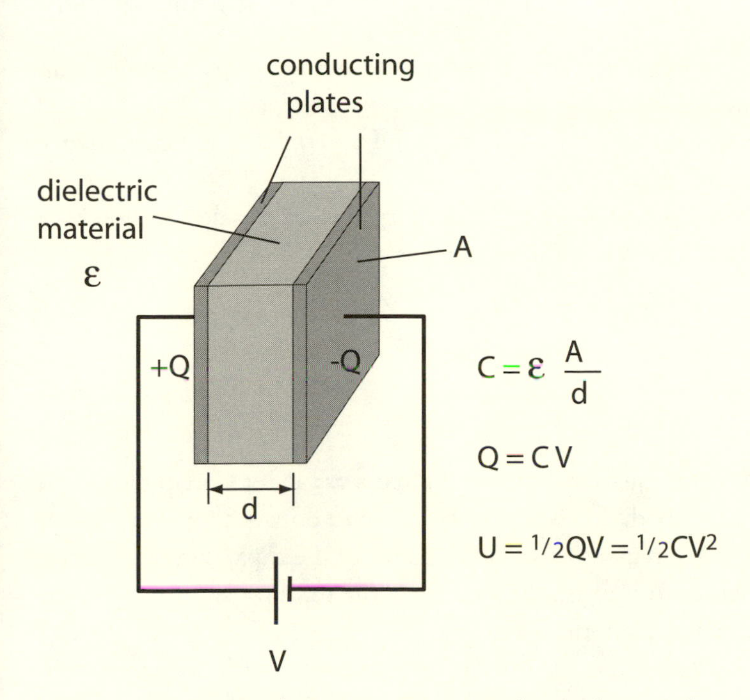

FIGURE 5.6. Properties of a capacitor. It stores electric charge ($+Q$ on one plate and $-Q$ on the other plate) and energy U. The amount of charge and energy stored depends on the capacitance C of the capacitor, which is determined by the physical size (area A and thickness d) and the dielectric permittivity ε of the material between the plates. The charge and energy also depend on the voltage V across the capacitor.

the voltage V of the battery to which it is connected. The expression that illustrates this dependence is

$$Q = CV. \tag{8}$$

Thus we see that the larger the voltage applied to the capacitor, the larger the charge that can be stored.

For the human body, the charge on either side of the cell membrane is not produced by connecting a battery, but involves passive processes such as simple diffusion through the membrane, binding to protein carriers that diffuse through the membrane, moving through protein channels in the membrane, or active processes that use energy (from adenosine triphosphate, or ATP) such as charge transport proteins embedded in the membrane. An example of a transporter is the sodium/potassium pump (more about this later). The amount of charge involved can be calculated by measuring the membrane potential and determining the capacitance of the cell membrane.

The electric potential energy of a charged capacitor depends on how much charge is stored and the voltage, as expressed in the formula

$$U = \tfrac{1}{2}QV \tag{9a}$$

or

$$U = \tfrac{1}{2}CV^2. \tag{9b}$$

Thus we see from equations (8) and (9) that a capacitor stores charge and energy. The energy that is stored is available to do work on charges that are present in the region between the plates of the capacitor, or if conditions are right, it can cause the stored charge to move so as to perform work or accomplish another task.

As an example, consider the flash on a camera. Batteries are used to charge a capacitor in the camera. When the button is pushed to take a picture, charge from the capacitor moves through the light bulb of the flash causing it to shine brightly for a brief moment. The energy stored in the capacitor is converted into light and heat emitted by the flash. Or think of flashing lights on orange barrels used in construction zones to help control the flow of traffic. Another example is an electronic keyboard: if a key is pushed, the distance between the plates of the capacitor changes, causing the capacitance to change and resulting in movement of stored charge.

Current and Resistance

Two other items to discuss before we look more closely at the human body deal with the amount of charge that moves, which is called *current*, and a property that has to do with limiting the flow of charge, called *resistance*. Current is defined as the amount of charge that flows past a point or through a wire divided by the time is takes to flow. Mathematically, the current I is written

$$I = \frac{q}{t}. \tag{10}$$

The charge can be electrons, protons, and/or ions. In most electrical circuits, it is electrons that move through wires. In certain special physics experiments, beams of protons move through vacuum chambers or tunnels. In the body, ions move across cell membranes. The unit for current is the *ampere*, or *amp*, abbreviated by A.

Often we find that the larger the electronic device, the larger the current required to run it. For example, calculators and MP3 devices can run on currents of a few milliamperes or less. Computers and televisions run on currents of a few amperes or so, whereas large objects like refrigerators or cars need tens of amperes or more to run.

Resistance is a property of a device that limits current flow in an electronic circuit. The device is called a *resistor*. Resistors can serve several purposes—they can be used (1) to limit current, or (2) to change an electric potential, or (3) to convert electric potential energy into work, heat, or light. The unit for resistance is the *ohm*, or Ω (Greek capital omega).

The value of the resistance depends on how a device is made (similar to a capacitor): the material that it is made from, its length L, and its cross-sectional area A (fig. 5.7). The expression that relates these parameters is

$$R = \rho \frac{L}{A}, \tag{11}$$

where ρ is the *resistivity* of the material. The resistivity is a property of the material related to another property, the *conductivity* σ, which indicates how well the material conducts electricity: $\rho = 1/\sigma$. Metals have a high conductivity, so they have a low resistivity. Electrical insulators, such as most plastics, are very poor conductors, so they have a high resistivity. Semiconductors have a moderate ability to conduct electricity; they can be made into conductors if certain atoms are added to them (a process called doping). Resistors are usually made of moderately conducting materials, although high-precision resistors are usually made of very long pieces of wire.

The current I that flows in a wire depends on the voltage V applied to the wire (or the potential difference along the wire) and the resistance R according to the expression

$$I = \frac{V}{R}. \tag{12}$$

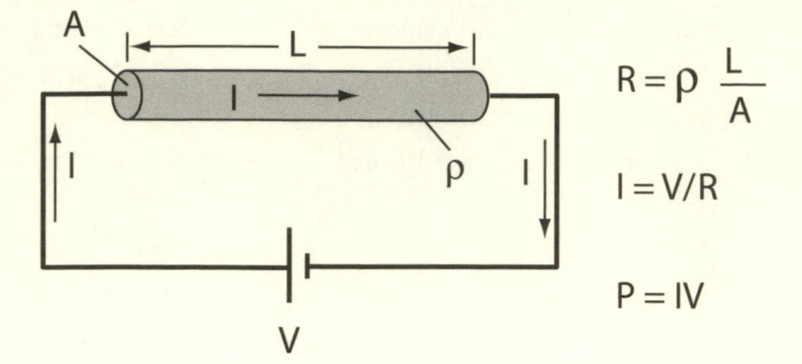

FIGURE 5.7. Properties of a resistor. It allows a certain amount of current I to flow in a circuit when connected to a battery of voltage V. The resistance R depends on the resistivity ρ of the material it is made from, as well as physical dimensions such as length L and cross-sectional area A.

In other words, in order to cause current to flow in a wire, a potential difference must be applied, which results in a force on the electrons of the metal in the wire, causing them to move. The electrons experience some resistance to their flow because of interactions with each other and with the underlying atomic structure of the material. Equation (12) is referred to as Ohm's law.

When charges move through an electric potential, energy is required, which we have seen can be determined using the expression $U = qV$. However, when we talk about current flow, it is the rate of energy usage or power that is important. Recall from chapter 3 that power is defined as the rate at which work is done, or the rate at which energy is converted from one form to another. Electric power is defined as the rate at which electric energy is converted to another form, such as heat or light, or to work, as when an electrical motor is used. In addition, electric power is used to describe the rate at which electric energy is produced, as in a battery or a power plant.

The mathematical formula that expresses electric power can be derived from the electric energy formula $U = qV$. If both sides of this formula are divided by time t, then U/t is power P and q/t is current I. Thus, we see that

$$P = IV. \qquad (13)$$

This expression says that the power supplied by a power source equals the current supplied by that source times the electric potential of the source. For example, suppose a car battery, which operates at 12 V, supplies 60 A when starting a car; the power supplied is 720 W.

Another interpretation of equation (13) is that the power used by an electrical device equals the current that flows through the device times the electric potential applied to it. As an example, the label on the back of my calculator says that it operates at 3 V and 60 μA. Thus, it uses energy at the rate of 180 μW, or 0.18 mW, or 0.00018 W.

Electricity in the Body

Now that we have looked at some of the basic principles of electricity, we can consider the human body. We will look at static electricity, current flow and skin resistance, the nervous system, and membrane potential, to name just a few topics.

Static Electricity

Static electricity refers to the buildup of excess charge on an object. "Static" is a word that seems to indicate that the charges are at rest, which is not necessarily true. Sometimes they are at rest, as when two pieces of clothing are pulled from a clothes dryer and they are stuck together ("static cling"). However, when you scuff your feet across the carpet, you build up static electricity on your body and the charges are moving during this process. So we should really just think of static electricity as an imbalance of charge on an object.

The excess charge can be generated in many ways, but usually is a result of moving two objects across each other. Scuffing your feet on the carpet, scooting across the seat of your car when you get out, rubbing a balloon in your hair, and holding onto a plastic railing as you walk down a set of stairs, like clothes brushing against each other in the dryer, generate static electricity. The buildup of charge usually works best on days when the humidity is low and when materials such as wool, fleece, rubber, and synthetic fibers are involved. All of these types of material, as you may have figured out, are insulators, so they don't conduct electricity easily. The transfer of charge to your body occurs when you are in contact with them.

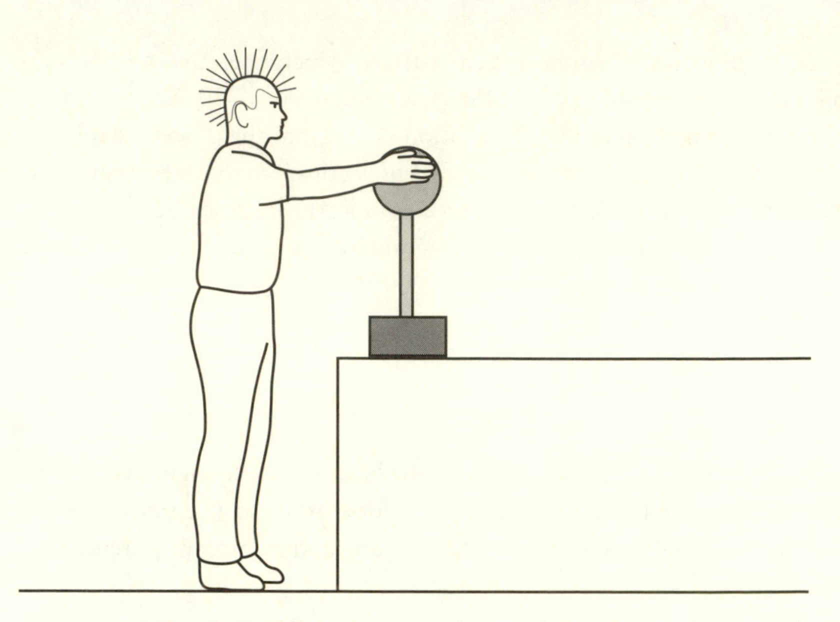

FIGURE 5.8. A Van de Graaff generator produces voltages up to 100,000 V. If a person touches the dome, charge is transferred to the person.

The excess charge on your body can be large enough that your body has a very high electric potential (voltage) up to 10,000 V or more. Perhaps you have seen demonstrations of a Van de Graaff generator where a person stands on an insulating block and places his or her hand on a large metal dome. The generator is turned on, causing a moving belt to brush against a metal wire that transfers charge to the dome and eventually to the person (fig. 5.8). Voltages as high as 100,000 V can be produced.

The charge gets distributed over the entire body so that if any part of you touches a metal object, such as a doorknob, you feel a shock. It is not the act of touching the doorknob that causes the shock, but the flow of charge through the air (the discharge) as it goes from you (at a high voltage) to the doorknob (at zero or very low voltage). The shock can be quite painful, so perhaps you get nervous or scared as you reach for the doorknob. I get shocked so often when I get out of my car that I often hit the door frame with my hand because the pain from hitting the door is a known feeling and occurs when I want it to; it tends to mask the pain (or the anticipation of pain) that comes from the unpredictable electric shock.

The excess charge on conductors tends to reside on the outside surface of the object. The human body can be thought of as a conductor when this idea of static electricity or charge imbalance is considered. Thus the excess charge on the body resides on the skin or hair. You may have seen someone with long hair in contact with the Van de Graaff generator, with hair sticking straight out from the head. The excess charge on the head tends to accumulate on the hair; the charges are most likely electrons so they repel each other, causing the hair to stand out similarly to electric field lines around a charge as in figure 5.3.

The Body as a Conductor

I mentioned above that the body is a conductor when it distributes excess charge through the body, but what happens if you touch an electrical wire? Will you get shocked? It depends on the electric potential or voltage of the wire. It will also depend on whether your skin is wet or dry. The body is not as good a conductor as metals such as copper and aluminum, but it is better than plastic and rubber. The reason that the body can conduct electricity is that water in the body contains many ions (Na^+, K^+, Cl^-, etc.) that will move when subjected to electric fields. Pure, deionized water does not conduct electricity.

Let's get a feel for some of the numbers. Suppose you hold your fingers across a 1.5 V flashlight battery. In this case you will not receive a shock, although there will be a small current flowing through your fingers. Typical resistance values between the fingers of the hand are several thousand ohms, giving a current from Ohm's law of less than a milliampere. However, if you were to run wires from the battery to your tongue, you would likely feel a tingling sensation because there is much more current flowing. This greater current arises because moisture on your tongue causes the resistance to be much lower. If you were to grab onto an electrical wire from a household appliance, where the voltage is about 120 V, the current would be much greater, causing severe injury or even death.

Most people can feel a current as small as 1 mA, with smaller currents being undetectable, and larger ones causing different reactions. Muscle contraction and loss of control occur when the current is about 10 to 20 mA, with the onset of pain at 50 mA. Higher currents can cause fibrillation

TABLE 5.1. *Effects of electrical shock as a function of current*

Current (mA)	Effect
1	Threshold of sensation
5	Maximum harmless current
10 to 20	Onset of sustained muscular contraction; cannot let go for duration of shock; contraction of chest muscles may stop breathing during shock
50	Onset of pain
100 to 300+	Ventricular fibrillation possible; often fatal
300	Onset of burns depending on concentration of current
6000 (6 A)	Onset of sustained ventricular contraction and respiratory paralysis; both cease when shock ends; heartbeat may return to normal; used to defibrillate the heart

Source: Paul Peter Urone. 2001. *College Physics,* 2nd ed. 492–493. Pacific Grove, CA: Brooks/Cole.

Note: For an average male shocked through the trunk of the body for 1 s by 60 Hz AC. Values for females are 60–80% of those listed.

of the heart and can be fatal. Table 5.1 lists effects of electrical shock through the trunk of the body that last for a period of 1 s.[1]

These levels of current passing through parts of the body depend on resistance and voltage when Ohm's law is applied: $I = V/R$. Dry skin has resistance values of 1000 to 100,000 Ω, while the resistance inside the body ranges from about 300 to 1000 Ω. Common values from hand to hand, foot to foot, and hand to foot range from 500 to 1500 Ω. All these values depend on the length, the cross-sectional area, and the conductivity of the types and layers of tissue (skin, muscle, bone, and so on) through which current travels. I remember in an electronics course that I took in college being told to use only one hand when reaching into an electronic device rather than two hands. If I were to be shocked, the current would go through only my hand in the device, and not through the hand and across my chest (very near my heart) and out the other hand. In fact, my teacher in that class showed us his scars from being burned when current went through his index finger to his thumb.

Nervous System

We have just seen that the body reacts in a variety of ways to external sources of electricity such as charge imbalance (static electricity) and electric shock. Internally, there is an amazing collection of electric charges that react to various types of external and internal stimuli. Their response helps the body to control many different kinds of activities, including motion, sensation, regulation, thought, and emotion. The control mechanism for these activities is called the nervous system.

The nervous system consists of nerve cells, or neurons, that are important for control and communication functions. There are three main uses of the nervous system. First, input to the system from its surroundings (external stimuli) and from other organ systems (internal stimuli) is sensed or detected; this function is called *sensory input*. Second, the nervous system processes this input information to interpret its meaning and to determine what to do in response to the information; this function is called *integration*. Third, the nervous system causes the body to respond; this is called *motor output*. Thus the nervous system receives input, processes the input, and produces a response.[2] For example, if you place your hand on a hot burner on the stove, you sense the heat, your nervous system processes the information (hot stove means remove your hand), and you quickly remove your hand.

The nervous system is organized into two main parts: the *central nervous system*, or CNS, and the *peripheral nervous system*, or PNS (fig. 5.9). The components of the central nervous system are the brain and the spinal cord. All other parts make up the peripheral nervous system that goes to other areas such as muscles, skin, the heart, and glands. The peripheral nervous system consists primarily of nerves for sensory input and motor output that communicate with the central nervous system (spinal cord and brain). It is important to distinguish these two functions of the peripheral nervous system. Sensory input is detected by sensory receptors; sensory nerve fibers carry the signal to the central nervous system. The central nervous system processes and interprets the input signal, decides what should be done, and produces an output signal. This output signal is transmitted along motor nerve fibers to effector organs, which are muscles that contract or glands that secrete. Some movement by skeletal muscles is voluntary, such as moving your legs when you walk. Other movement is involuntary, such as the beating of the heart or the movement of food through the digestive system. Secretion of glands is also involuntary.

The nervous system consists of two types of nerve tissue: *neurons* and support cells (*neuroglia* or *glial* cells). Neurons are nerve cells that

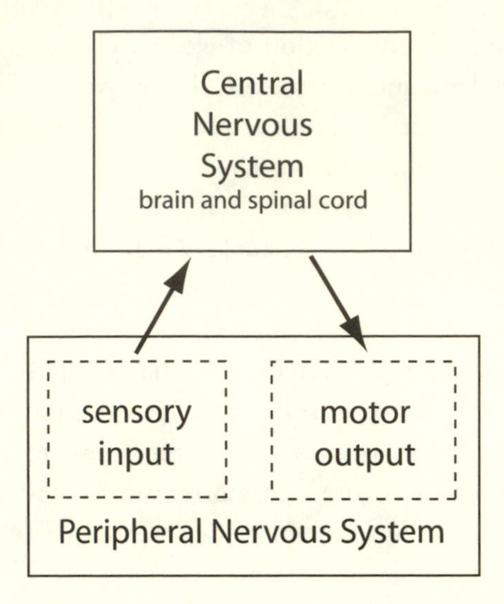

FIGURE 5.9. The human nervous system. It consists of the central nervous system and the peripheral nervous system. Sensory nerve fibers transmit electrical signals to the central nervous system, which interprets the information and sends signals to motor nerve fibers to produce a response.

transmit the electrical signals from the sensory receptors to the central nervous system and from the central nervous system to the effector organs. Neuroglia provide structure, support, and protection for the neurons; they also help make connections and provide insulation for signal transmission in the neurons. Certain types of neuroglia wrap tightly around neuron fibers to produce what are called *myelin sheaths*.

Neurons

Neurons, or nerve cells, are large cells that consist of a cell body and two types of extensions called *dendrites* and *axons*. The cell body contains the nucleus of the cell and other components that make proteins, maintain the cell membrane, and control the flow of electric signals through the neuron. Dendrites are short, thin, multibranching tendrils connected to the cell body that receive input signals from other neurons and carry the signals to the cell body. Each neuron has only one axon, which is a long tendril extending from the cell body that carries electrical signals to its end. The terminal end of an axon may have several branches, called nerve endings or axon terminals (fig. 5.10). When these terminals receive electrical signals, they release chemicals called neurotransmitters to communicate with other neurons or cells. The connections between neurons, or between neurons and other cells, are called *synapses*.

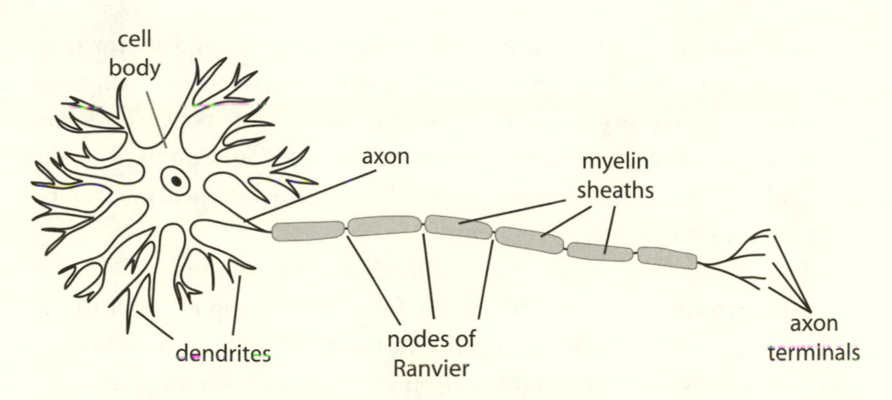

FIGURE 5.10. A nerve cell or neuron consisting of a cell body, dendrites, and an axon. Electrical signals are detected by the dendrites and sent along the axon to the axon terminals. The terminals pass the signal to other neurons or to effector organs.

Neurons are classified according to their structure. The structural classes include multipolar, bipolar, and unipolar neurons. Multipolar neurons have many dendrites and one axon extending from the cell body. Bipolar neurons have only one dendrite (with many branches) and one axon extending from the cell body. Recall that dendrites are the receptor regions of the neuron and the axon terminals are the secretory regions. Unipolar neurons have only an axon extending from the cell body; one end of the unipolar neuron is the receptor region (which is actually sometimes called the dendrites) and the other end is the secretory region.

Neurons are also classified according to their function. The functional classes include sensory neurons, motor neurons, and interneurons.[3] Sensory neurons conduct input signals toward the central nervous system. Almost all sensory neurons are unipolar, with the receptor region located in the peripheral nervous system, such as in the skin to detect touch or temperature. Bipolar neurons are sensory neurons as well, although they are rare, and are located in special senses, such as the eye, the nose, and the ear.

Motor neurons conduct electrical signals from the central nervous system to effector organs (muscles and glands). They are multipolar neurons with their cell bodies located in the central nervous system and their axons located in the peripheral nervous system, such as in a muscle of the arm to move it quickly in response to touching something hot.

Interneurons are situated between sensory neurons and motor neurons, either part of a chain or as a single neuron. Interneurons are almost entirely multipolar neurons. Thus sensory input to the dendrites produces an electronic signal that flows toward the cell body and onward to the axon. More than 99% of the neurons of the body are interneurons, which reside in the central nervous system.[4]

The five senses—touch, taste, smell, vision, and hearing—all involve sensory neurons. The electrical signals from sensory input travel to the axon nerve endings (secretory region) located in the central nervous system where they pass their information along to the interneurons. The signals continue to other neurons, either to interneurons in a chain or to motor neurons. The signals ultimately end up at the cells of effector organs.

Bundles of axons in the peripheral system are called nerves. They can be quite long; for example, if the cell body is located near or in the spinal cord, then axons as long as 1 m or more are required to reach all the way to the foot. A pinprick on your big toe sends a signal through the long axon of a sensory neuron to the central nervous system, through an interneuron, to a motor neuron, continuing to its long axon, and to a muscle so that you can move your big toe away from the pin.

An important physical characteristic of axons is that they are wrapped with myelin sheaths, which are actually cells (Schwann cells). Each sheath is only about 1 mm long so that there are many of these sheaths along the length of each axon. There is a gap of about 1 μm between the sheaths; these spaces are called *nodes of Ranvier* (see fig. 5.10). The sheaths and the nodes affect the ability of an axon to conduct the electrical signals.

Electrical Properties of the Neuron

What are these electrical signals and how do they move through the neurons? How does all this relate to the various electrical concepts that were developed earlier? We now explore the electrical properties of the neuron, including the movement of charge and the presence of electric potential in the cell. We will also see that some chemistry concepts, such as diffusion and molecular transport, play important roles as well.

The movement of electrical signals along an axon is not like the movement of current in a wire in a household appliance. Current in wires consists mostly of electrons that move collectively along the direction of the wire. Electrical signals in the body involve ions that move across the cell membrane perpendicular to the axon in response to a stimulus applied at a certain point. The perpendicular motion of these localized ions produces a change in the electric potential that causes nearby ions to pass through the membrane. There is some movement of charge parallel to the axon, but without the perpendicular motion the parallel motion would die out quickly. The process continues along the length of the axon similarly to a wave or pulse moving along a rope or a spring.

Membrane Potential

A cell has a number of different components, but we will focus on the charges distributed through the fluid on the inside and on the outside of the cell, as well as movement of charges through the cell membrane that separates these two regions. The two fluids (inside and outside) are both electrically neutral, but they contain many different kinds of ions (fig. 5.11). The cell interior tends to have a much higher concentration of potassium ions (K^+) than exists outside the cell. These positive ions are electrically balanced by large, negatively charged proteins (A^-). Outside the cell, there is a much higher concentration of sodium ions (Na^+) than inside. These positive ions are electrically balanced by negatively charged chlorine ions (Cl^-) and smaller quantities of other anions. There are other ions

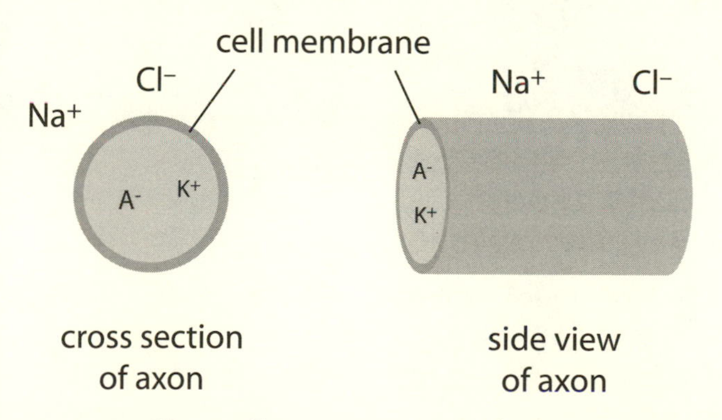

FIGURE 5.11. Nerve cells contain many ions both inside and outside the cell. Examples are sodium Na^+, chlorine Cl^-, potassium K^+, and other negative ions represented by A^-. Overall the cells are electrically neutral, but the ions can respond to electrical forces and other forces.

and molecules in the fluids, but the sodium and potassium ions play the most important role in determining the electrical properties of the cell.

The membrane has different degrees of permeability for these different types of ions. Permeability has to do with the ability of a particle to move through the membrane. The membrane is usually impermeable to the large, negatively charged proteins, so they cannot move through the membrane—this keeps their concentration constant inside the cell. The membrane is slightly permeable to the sodium ions, much more permeable to the potassium ions, and nearly completely permeable to the chlorine ions.

Because of the higher concentrations of K^+ ions inside the cell and Na^+ ions outside the cell, there is a tendency of these ions to move through the cell membrane. This tendency is caused by a process called *diffusion*: particles tend to move (diffuse) from regions of higher concentration to regions of lower concentration. The difference in concentration is called a concentration gradient (or chemical gradient). If you were to remove the top from a bottle of perfume, the aroma would eventually fill the entire room because of diffusion. Based on the diffusion process, potassium ions prefer to move out of the cell and sodium ions try to move into the cell (fig. 5.12). However, because of the different degrees of permeability to

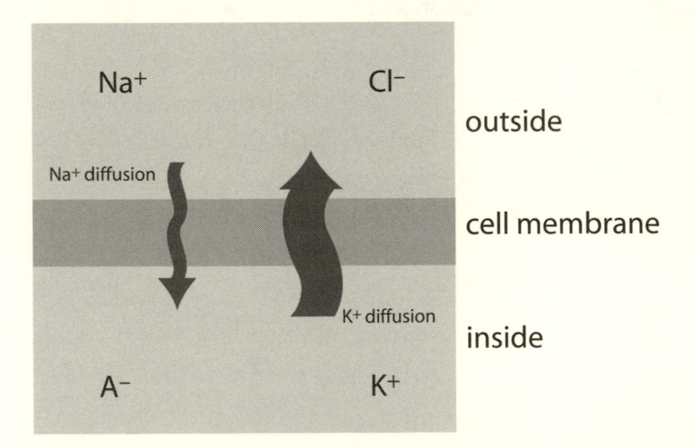

FIGURE 5.12. Diffusion of ions across the cell membrane caused by concentration gradients. The permeability of the cell membrane is different for sodium Na^+ and potassium K^+ ions. Potassium ions move out of the cell at a greater rate than sodium ions move into the cell.

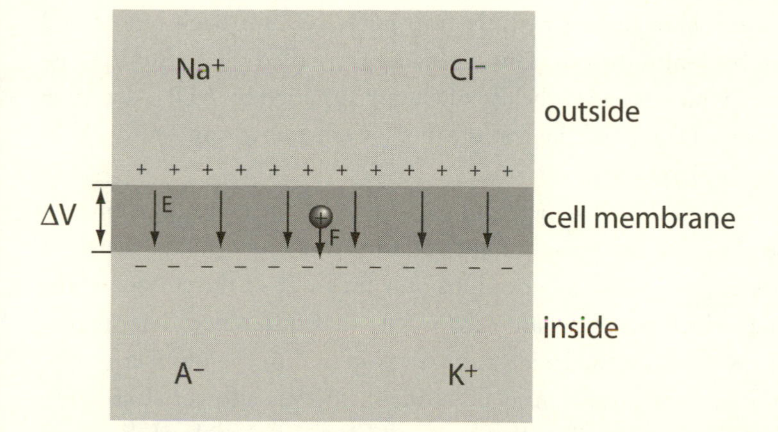

FIGURE 5.13. The membrane potential. In the resting state, there are more positive charges just outside the cell membrane than just inside. This difference in number of ions causes an electric potential difference ΔV, called the membrane potential, to exist across the membrane; the electric field E within the membrane points toward the inside of the cell. A positive charge would tend to move toward the inside of the cell because of the electric force F acting on it.

the various ions there is a greater tendency of K^+ ions to move out of the cell than of Na^+ to move into the cell. When more positive ions move outward than inward, the inside of the membrane becomes more electrically negative than the outside. Thus, an electric potential difference develops across the membrane (fig. 5.13).

The presence of this membrane potential causes another force, an electric force, to act on the ions. Positive ions tend to move from higher electric potential to lower potential—this is called an electrical gradient. The electrical gradient would make both K^+ and Na^+ ions want to move from outside to inside the cell. This direction is the same as the concentration gradient for sodium, but opposite to it for potassium.

The flow of ions across the cell membrane occurs as a result of these two gradients: the concentration gradient and the electrical gradient. These two components to the motion of ions are passive transport mechanisms because they require no input of energy. If the cell were left alone, the concentrations of the ions would eventually equalize inside and outside the cell. However, because the concentrations of sodium ions inside the cell and potassium ions outside the cell are quite constant there must be a mechanism that actively transports sodium back out of and potassium back into the cell. This mechanism is called the sodium/potassium pump, which requires the use of energy supplied by ATP (adenosine triphosphate). Details of the sodium/potassium pump can be found in textbooks on physiology.[5]

As a result of the movement of ions caused by the concentration gradient, the electrical gradient, and the sodium/potassium pump, there is a slight difference in the number of ions on either side of the surface of the membrane, which produces an electric potential difference. The smaller number of K^+ ions on the inside surface compared to the number of Na^+ ions on the outside surface means that the inside has a lower electric potential than the outside. The electric potential of the outside of the membrane is normally taken as the reference point (or zero) for the potential, which means that the potential inside is negative. In the equilibrium state or the resting state, the membrane potential is about –70 mV, although it varies for different types of cells from about –40 to –90 mV.

Action Potential

When the cell is in the equilibrium state and the potential difference of –70 mV exists (the resting membrane potential), the membrane is said to be polarized. Under a variety of conditions, charges can move into or out of the cell causing changes to the membrane potential. If the potential difference becomes less negative (moves toward 0 mV) or becomes positive, the cell is said to be depolarized. If it becomes more negative (than –70 mV), it is said to be hyperpolarized.

Changes to the membrane potential occur as ions move across the cell membrane along certain routes called *channels*. These channels are proteins that help with ion movement when they are open and block ion movement when they are closed. Some channels are gated, which means they can be opened or closed when certain chemicals bind to the protein (chemically gated or ligand gated), when the membrane potential is at a certain value (voltage gated), or when the membrane is physically deformed (mechanically gated). Other channels are always open such as those used for diffusion. The channels are selective in that they allow only a certain type of ion to cross the membrane.

When a stimulus is applied to a neuron, the cell is no longer in the equilibrium state and gated channels will open. This action causes (allows) ions to move across the membrane, which changes the membrane potential. The size of the change in potential depends on the strength of the stimulus.

Graded potentials are localized changes in the membrane potential that depend directly on the strength of the stimulus. The size and lifetime of graded potentials also depend on the strength of the stimulus. When a stimulus is applied at a small region of the membrane certain gated channels open up, causing ions to move across the membrane. In that region, the membrane becomes depolarized (the membrane potential is more positive than the resting potential). This region of depolarization affects neighboring ions in the polarized (resting) region, causing positive ions to move toward these newly formed regions of lower potential and negative ions to move to the newly formed regions of higher potential. These currents are produced along the length of the neuron, parallel to the membrane.

However, while the parallel motion is going on, charges diffuse back through the cell membrane perpendicular to the membrane. This diffusion, or leakage, produces what is called a capacitance current because you can think of the charge separation across the membrane as a charged capacitor. The material filling the space is not a perfect insulator, so the membrane can be thought of as a charged capacitor that temporarily stores charge. Even though the neighboring charges are affected and current flows parallel to the membrane, there is so much leakage across the membrane that the resting potential is reestablished very quickly and the distance along the axon that is affected is quite small.

Action potentials are much stronger than graded potentials. When a large enough stimulus is applied, a large depolarization occurs causing the membrane potential to change from –70 mV to +30 mV (fig. 5.14). This event is followed by repolarization, including a slight hyperpolarization (more negative than the resting potential), before the membrane returns to the resting (polarized) state. The time for these changes to occur is only a few milliseconds. However, this effect is so strong that this action potential, or nerve impulse, is propagated along the entire length of the nerve.

Not all stimuli produce an action potential. If the stimulus causes a depolarization of at least 15 to 20 mV above the resting potential, which

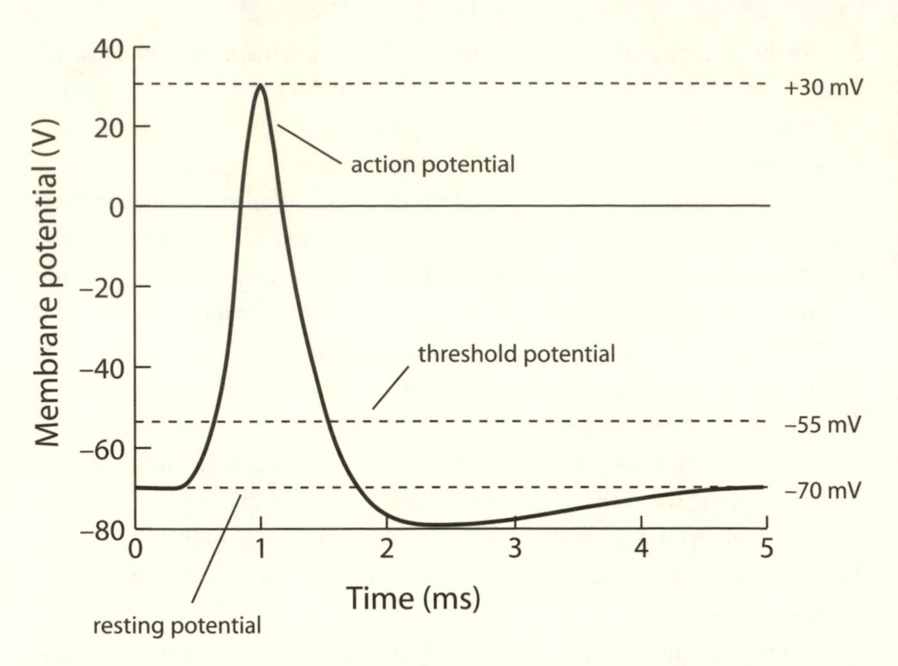

FIGURE 5.14. Action potentials. When channels open in the membrane because of a stimulus, ions can move quickly across the membrane, changing the polarity of the potential. When the potential changes from −70 mV to +30 mV, the process is called depolarization. When it returns to the resting condition, it is called repolarization. When the potential overshoots the resting potential during repolarization, the process is called hyperpolarization.

is about −55 to −50 mV and is called the *threshold potential*, then the action potential is generated. For smaller values of depolarization a graded potential is the result.

Keep in mind that a stimulus causes some of the gated channels to open. In particular, voltage-gated channels that allow Na^+ to enter begin to depolarize the region where the stimulus occurs. If enough sodium ions flow in and the threshold value of the membrane potential is exceeded, more and more sodium channels open up, allowing sodium to pour into the cell. This results in the membrane potential shooting past 0 mV and going as high as +30 mV. At this point, the sodium channels begin to close, but K^+ channels begin to open allowing potassium ions to flow out of the cell—this is called the repolarizing phase. This phase takes place quickly as well, with the polarization moving slightly past the resting potential of –70 mV resulting in hyperpolarization.

Even though the proper sign of the polarization has been reestablished, the concentrations of the sodium and potassium ions are reversed. It is the sodium/potassium pump that works to return these ions to their proper locations—sodium ions outside and potassium ions inside the membrane.

As stated previously, this entire sequence of events takes only a few milliseconds to occur. The ions move quickly, currents start and stop quickly, and the cell returns to its resting state in a very short period of time. Remember, however, that these changes take place over a small region of space where the stimulus occurred.

It is this action potential that is propagated along the axon. The action potential causes ions in regions nearby to move, causes ion channels to open and close in nearby regions, and results in a positive feedback mechanism that keeps the action potential moving away from the original site.

The speed of the propagation depends on whether the axon is myelinated or unmyelinated. In unmyelinated axons, the propagation is rather slow, because conduction has to occur across continuous regions of the membrane. They are depolarized and repolarized similarly to a wave traveling as a pulse on a string. For myelinated axons, the myelin acts as an insulator and keeps the leakage of charge (from diffusion) low. The effect is that the current parallel to the axon moves much more quickly. Movement of charge perpendicular to the axon occurs only at the nodes of Ranvier where there is no myelin. The action potential is once again generated at a node, and the effect on the neighboring myelinated region is to produce a fast-moving current to the next node. Nodes are separated by about 1 mm. A new action potential is generated at each node, and the electrical signal moves quickly from node to node along the axon. This type of conduction is called *saltatory conduction*.[6] It occurs at speeds of 100 m/s or more, much faster than continuous conduction in unmyelinated axons, which occurs at speeds of only a few meters per second.

Once an action potential is produced, it is independent of the strength of the stimulus. However, some stimuli are larger than others. For example, a painful pinprick on the bottom of your foot or a single hair sweeping across your face will elicit very different responses. If a stimulus is particularly strong, then the action potential is generated more often. The central nervous system interprets this high frequency as an intense stimulus and responds appropriately.

When an action potential is generated and the ion channels are open, there is a period of time during which another stimulus at that location in the neuron will not cause much, if any, response. During depolarization when the sodium channels are open until they start to close, no response to a stimulus occurs; this period of time is called the *absolute refractory period*. During repolarization, as the sodium channels are closing and the potassium channels are open, weaker stimuli will not cause a response, but very strong stimuli can produce another action potential; this period of time is called the *relative refractory period*.

Other electrical properties of the axon are important but are not discussed much more here. The capacitance of the axon can be used to describe the amount of charge stored on either side of the membrane and the energy stored. The resistance of the axon allows us to discuss the current that flows perpendicular and parallel to the axon as well as the power used in these processes. For more details on some of these topics please refer to the references listed.[7,8,9]

Interactions of Signals

I have presented these ideas as if a single stimulus produces an action potential in a single neuron. However, neurons work together in groups, which work together in even larger groups. There are often multiple receptor sites, whose neurons are interconnected with other neurons, branching out in a diverging set of pathways. Then there are signals that converge from large bundles or from several regions to fewer and fewer neurons.

There are millions of neurons in the central nervous system that integrate all the input information in order to process and pass on a response to other areas. The input information can come from nerves in the peripheral nervous system, or from other neurons within the central nervous system. The output is passed on to other nerves in the central nervous system or to nerves in the peripheral nervous system.

Have you ever seen a network of computers, servers, and other electrical equipment together in a large IT (informational technology) room?

Or have you thought about what the internet looks like? Individual computers (with their individual components) are wired together all over the world. They each perform their own tasks, but groups of computers perform other tasks, and groups of groups of computers perform other tasks as well. They work sequentially (*serial processing*) or side by side (*parallel processing*).

The nervous system is very similar to the network of computers. Serial processing and parallel processing happen constantly. Reflexes are an example of serial processing where a rapid response to a stimulus occurs, such as pulling your hand away from the hot stove. Thinking, emotion, and recall are examples of parallel processing where multiple responses arise as a result of a stimulus. Suppose you see an apple pie in an old country store; it might remind you of your childhood and of how much you enjoy apple pie, and you might think about going apple picking over the weekend.

Applications

I have spent a quite a bit of time talking about the nervous system and nerve conduction, but there are other electronic applications related to the human body. By applying basic electrical principles, we can have a better understanding of how certain electronic devices work. Specifically, I want to briefly describe the *electrocardiogram* (ECG or EKG), the *electroencephalogram* (EEG), the *electromyogram* (EMG), and the *defibrillator*.

The EKG is a recording of the electrical signals that arise because of the electrical activity of the heart. As the heart goes through its cycle of pumping blood, the electrical impulses that trigger the heart to contract and relax also cause depolarization and repolarization of the nerves that go to the heart. The changing membrane potential in these nerves can actually be measured at some distance from them because of the electric fields produced. These electric fields act on electrical probes placed on the body. In fact, when an EKG is measured there are typically 12 electrical leads attached to the arms and legs and across the chest. These leads are able to measure potential differences across different parts of the body as a function of time. When these signals are combined, an EKG is produced.

There are four chambers of the heart, the left and right atria (the upper chambers) and the left and right ventricles (the lower chambers). The sequence in which they individually contract and relax is important for proper flow of blood. The EKG measures the timing of electrical impulses that control the sequence of the heart.

An example of a normal EKG is shown in figure 5.15. There are several distinguishing features. Briefly, the P wave arises when the nerves depolarize causing the right and left atria to contract. The QRS portion corresponds to depolarization of the nerves to the ventricles followed by their contraction. The T wave indicates repolarization of the nerves to the ventricles after which the heart relaxes. Repolarization of the atria is obscured by the QRS portion of the recording. There are other features in an EKG that are observed occasionally, such as a U wave that follows the T wave; it represents the repolarization of the *Purkinje fibers* that help coordinate the timing of ventricular contractions.

The electroencephalogram also measures electrical activity in the body. An EEG is measured by placing many electrodes on the scalp; these electrodes are able to detect the electric fields produced by moving charges in the synapses (connections) between neurons in the brain. The patterns that are detected and recorded are called brain waves. There are several types of brain waves: alpha waves that correspond to an awake, relaxed state; beta waves that correspond to an alert state; theta waves that are common in children, but rare in adults; and delta waves that occur during deep sleep.

An electromyogram is a measurement of the electrical activity in muscles. When the action potential in the axon reaches a muscle, it causes the muscle to contract. Usually there are many nerves (or their ends) attached to many muscle cells, which cause the muscle to contract so as to move a particular part of the body. Electrodes detect the electric signals that result from the stimulation of a muscle. The EMG can be studied to determine whether the muscle is functioning properly or not.

The defibrillator is a device used to establish the normal rhythm of the heart. Two plates, or paddles, are placed one on either side of the heart. A large capacitor is charged such that when a switch is pushed, the charge is dissipated through the body. The purpose is to electrically shock the heart and interrupt an abnormal rhythm, called *cardiac arrhythmia*. The shock should be strong enough to depolarize the nerves that go to certain points called *nodes*, in particular the *sinoatrial*, or *SA*, *node*. (It

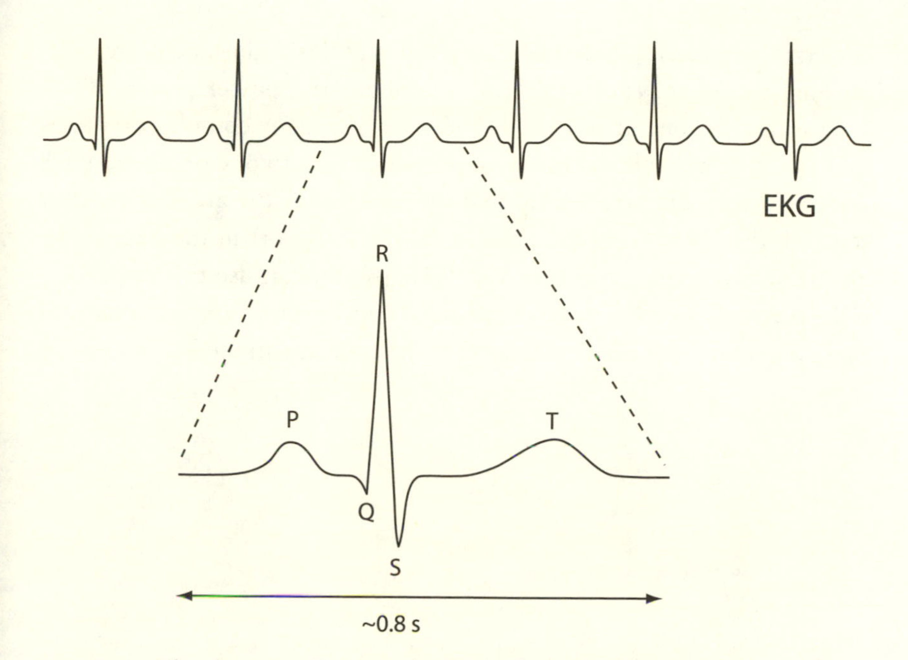

FIGURE 5.15. The electrocardiogram. An EKG of a normal heart pattern is detected by electrical leads attached to the body. Depolarization and repolarization of the nerves that go to the heart produce electric fields that are picked up by the leads. The point P on the graph corresponds to depolarization of the nerves that cause the two atria to contract. The QRS points correspond to depolarization of the nerves to the ventricles, causing them to contract. The point T corresponds to repolarization of the nerves to the ventricles.

is the SA node that controls the heart rate and rhythm.) It is hoped that this will restart the normal electrical sequence resulting in a return to the normal contraction and relaxation of the heart. Defibrillators are usually external devices, but for patients who have had past problems with cardiac arrhythmias, including ventricular tachycardia or ventricular fibrillation, an internal defibrillator may need to be implanted and attached to the heart.

Summary

This chapter on electrical properties of the human body has included a number of different physics concepts and applications. The physics discussed is used in simple electric circuits such as flashlights and more complicated electrical systems such as computers and power plants. When discussing the human body, we see that there are many complex processes that occur in a single neuron as well as in the network of the nervous system. A physiologist recently told me that one of the areas of greatest research these days is in the area of charge transport in the neuron, so all the answers are not known yet. Perhaps you can use this discussion to help you understand some of the basic concepts and applications, and then to explore in more detail some of the areas mentioned.

SIX

Optics of the Eye

One of the most important senses that humans have is the ability to see. Much of our perception of and interaction with the world around us involves sight. Within just moments after birth, babies begin to recognize faces and colors. From words on a page to paintings in a museum to the looks on people's faces, visual images have profound effects on how we learn, how we live, and how we respond.

Images are formed when light from an object passes into the eye and strikes the retina. We can describe this process using basic physics concepts from geometric optics, specifically the refraction of light and image formation by lenses. But how are we able to see letters on the page of a book, which is located quite close to the eye, and then look at a mountain range or the moon, which is located very far away? And why do so many people need contacts or glasses? What is LASIK, and how does it help to correct poor vision? These and other questions will be addressed in this chapter.

Electromagnetic Waves

Visible light is a type of *electromagnetic wave* that we humans can see. Electromagnetic waves, or EM waves, consist of electric fields and magnetic fields that vary as sine or cosine functions in space and time. They are produced by electric charges that move.

We discussed in chapter 5 the concept of electric fields produced by charges. A charge can exert an electric force on another charge if the second charge is placed anywhere around the first charge. If the second charge is removed, an electric field remains in the space around the first charge.

Magnetic fields are similar to electric fields. However, a charge produces an electric field no matter if it is stationary or if it is moving, but for a magnetic field to be produced, the charge must move. Magnets are made from materials where the electrons rotate around the nuclei of their associated atoms in a collective manner so that they produce a magnetic field throughout the material; the influence of this field extends outside the magnet itself. You are probably most familiar with magnetic fields if you have ever held two magnets near each other. You can feel the attractive and repulsive forces that act between the north and south poles of the magnets.

EM waves are produced in several ways. Radio waves are produced when electrons move back and forth in an antenna at a radio station. When electrons in an atom move from high- to low-energy states, they often emit visible light, ultraviolet radiation, or x-rays. The important point here is that the electrons are moving.

In an EM wave, the electric fields and magnetic fields oscillate back and forth exhibiting simple harmonic motion, as discussed in chapter 4 for sound waves. They have typical properties of waves including frequency, wavelength, amplitude, and speed.

There are several different types of EM waves whose primary difference is the frequency. All EM waves consist of electric fields and magnetic fields as described above. Also, they all move at the same speed in vacuum, called the speed of light, which is given by $c = 3 \times 10^8$ m/s = 300,000 km/s = 186,000 mi/s. (The c in this expression is the same c that is in Einstein's famous equation $E = mc^2$, which relates mass and energy.)

The different types of EM waves form what is called the *electromagnetic spectrum*. The electromagnetic spectrum has a range of frequencies from lowest to highest in the following order: radio and TV waves, microwaves, infrared (IR) radiation, visible light, ultraviolet (UV) radiation, x-rays, and gamma rays. Recall from the discussion of sound waves that the product of the frequency and the wavelength is the speed; thus if all EM waves move at the speed of light c, their wavelengths depend on their frequencies. The order written above for the different types of EM waves in terms of *increasing* frequency is the same order for *decreasing* wavelength. Radio and TV waves have the longest wavelengths, which are comparable in size to large antennae used for radio and TV stations; gamma rays have the shortest wavelengths, which are about the same size as the nucleus of an atom. The wavelengths of visible light are roughly the same size as molecules. In a chapter on optics of the eye, we will focus mostly on visible light.

Just as sound waves have energy, EM waves also have energy. The energy for sound waves in a gas is related to the amplitude of vibration of the gas molecules. For EM waves, the energy E depends on the frequency f of the wave, and is expressed mathematically as

$$E = hf, \tag{1}$$

where h is Planck's constant given by $h = 6.63 \times 10^{-34}$ J s. Thus, the order of the different types of waves (from radio and TV waves to gamma rays) is the same order as for energy: radio and TV waves have the lowest energy, then microwaves, IR radiation, visible light, UV radiation, and x-rays, and the highest-energy EM waves are gamma rays.

Visible Light

Visible light is the main concern for a chapter on optics of the eye. Visible light is part of the electromagnetic spectrum, but it also has a spectrum of its own. The spectrum has to do with color, written here in order of increasing frequency (or decreasing wavelength, or increasing energy): red, orange, yellow, green, blue, and violet. As stated previously, visible light moves at the speed of light given by $c = 3 \times 10^8$ m/s in vacuum. The range of wavelengths (in vacuum) is from about 700 nm for red light (just on the edge of infrared radiation) to about 400 nm for violet light (just on the edge of ultraviolet). Table 6.1 lists the various colors and their respective wavelength ranges. The frequency range for visible light is from 4.3×10^{14} Hz (red) to 7.5×10^{14} Hz (violet), and the energy range is from 1.8 to 3.1 eV.

Visible light is mostly produced by transitions of electrons as they move into or out of various energy states in atoms and molecules. When a light bulb glows, there are electronic transitions in the filament of a tungsten lamp or in the gas of a fluorescent lamp. Light from the sun is

TABLE 6.1. *Wavelength ranges for electromagnetic radiation*

Color or type	Wavelength range
Ultraviolet	<400 nm
Violet	400–440 nm
Blue	440–500 nm
Green	500–560 nm
Yellow	560–590 nm
Orange	590–625 nm
Red	625–700 nm
Infrared	>700 nm

Note: The wavelength range for visible light is from 400 to 700 nm. The boundaries between the different colors are not distinct; this means that each color transitions to the next color in a continuous fashion. For example, a wavelength of 500 nm corresponds to a blue-green color. Even the boundaries between UV, visible, and IR are a bit fuzzy, with some sources putting the range for visible light from 380 to 750 nm.

produced by electronic transitions that are caused by heat in chemical and nuclear reactions. Other types of EM waves are produced in these reactions but it is only the visible portion that we see.

The energy range for visible light turns out to be similar to the energies required to move electrons into and out of different energy states of atoms and molecules. In some cases, electrons are completely knocked out of an atomic or molecular orbital to become "free" electrons. This phenomenon is important for human vision because the retina of the eye has special types of receptors that respond to the absorption of different colors of light. Their response usually involves the excitation or removal of electrons from their molecules. These electrons produce a current that is detected by the optic nerve, which sends an electrical signal to the brain so that it can be interpreted as sight. We will look more closely at the anatomy and physiology of the eye later.

As we continue our discussion about visible light, I will typically drop the use of the word "visible." From now on, I will use the term "light" to refer to visible light.

Waves and Rays

Light is a wave with frequency and wavelength, as are EM waves in general. Waves have a variety of properties, some of which we discussed back in chapter 4. Light emanates from a source, such as the sun or a candle or a light bulb, moving away from it. Just as a point source of sound produces sound that emanates equally in all directions as if moving through an imaginary sphere, light can also be generated by point sources, such as a small light bulb or a small burning candle. The picture is that of a series of waves moving outward from the source like concentric circles (in two dimensions) or spheres (in three dimensions); these circles or spheres are called *wave fronts*. The distance between each of these wave fronts is the wavelength; the wave fronts move outward at the speed of light c (fig. 6.1a).

If we are somewhat close to the source of light, then the wave fronts look like circles or spheres. However, if we are very far from the object, then the wave fronts look like a series of parallel straight lines (in two dimensions) or parallel planes (in three dimensions). A way to think about this transition from spheres to planes is to consider what a small ball looks like compared to the huge earth. A small ball is highly curved and we can see its entire shape. The earth looks like a flat plane (when we are standing on it) and we can see only a relatively small portion of it. The center of the earth is about 6400 km (4000 miles) from the surface; if we

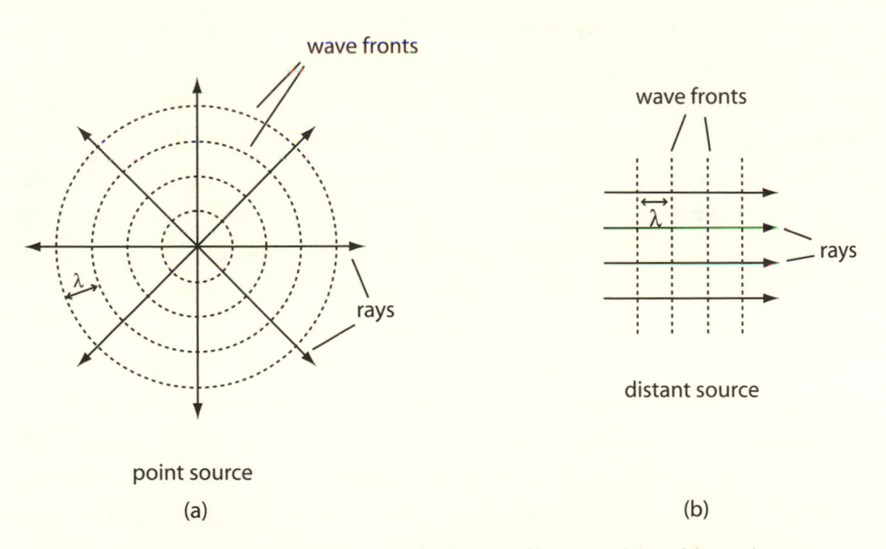

FIGURE 6.1. Light represented as a series of wave fronts separated by a distance equal to the wavelength λ. Rays are perpendicular to the wave fronts and point in the direction in which the wave fronts are moving.

had a point source of light 6400 km away, the wave fronts from it would look like planes to us (fig. 6.1b). A similar situation occurs with the sun: it is actually a very large source of light, but we are so far away from it that in many cases it can be considered a point source of light; the wave fronts that reach the earth are plane waves. This *wave picture* of light is useful for describing a number of different observable behaviors of light, such as diffraction and interference.

Another way of considering light is to use the *ray picture*. This focuses on the direction in which the wave is traveling: a ray is an arrow that points in the direction the wave is moving, which is perpendicular to the wave fronts. For example, the spherical wave fronts from a point source represent the wave picture, but rays pointing outward from the source (like the spokes on a bicycle wheel) represent the ray picture. The planar wave fronts from a source of light that is very far away represent the wave picture, but parallel rays pointing in the direction in which the waves are moving represent the ray picture.

You can think of a ray of light as a beam of light coming from a flashlight or a laser. It moves in a certain direction; it can illuminate objects; it can reflect off a mirror; and it can bend as it moves from, say, air to water. Rays of light can move through a variety of different materials as long as the material is transparent, such as air, water, glass, and so on.

Light rays move in straight lines through a material as long as the type or properties of the material does not change. If a light ray is traveling in air and strikes the surface of a lake, then it can reflect off its surface, or it can move into the water. If it moves into the water it bends, or refracts. *Reflection* and *refraction* of light are important concepts that happen all the time. We will explore refraction in more detail later.

Rays of light from a point source point outward from the center, but rays of light from a distant object appear to be parallel to each other. For example, if we stick needles into a small ball perpendicular to its surface, they will point radially outward from the ball. However, if we pound several long stakes into the ground perpendicular to the surface, they will appear to be parallel to each other. This concept of parallel light rays from a distant source of light is important when we discuss refraction of light in lenses and the eye.

As we continue in this chapter, it may be necessary to use the wave picture or the ray picture of light so it has been necessary to describe these two views. However, for most of the rest of the discussion we will use the ray picture. When we do, we will see that there are quite a few principles from geometry that arise. The study of light as it moves through different transparent materials, or transparent media, using the ray picture is called geometric optics. To understand how the eye works, we need to explore this topic more closely.

Geometric Optics

As mentioned previously, light rays follow very specific paths as they move through various media. The path of a light ray can be described using some basic concepts of geometric optics. The primary phenomenon is *refraction*, or bending, of light as it travels from one transparent medium into another.

Refraction is demonstrated easily in water; for example, when you look at a straw or a pencil half submerged in a glass of water, it may appear to be broken. A fish in an aquarium appears to be in one location, but is actually located at a different place. Light from the fish passes through the water, strikes the side of the aquarium, and refracts so that it moves in a different direction. If we stand in a clear pool of water and look down at our legs, they appear short and stubby because of refraction.

Refraction occurs in a variety of optical devices, such as magnifying glasses, telescopes, and microscopes. Cameras, binoculars, and eyeglasses are other examples. Each of these instruments causes light to refract in specific ways to form images that can be photographed or viewed with the eye. Even the eye itself can be thought of as an optical device. (Mirrors are also optical devices but they depend on the reflection of light rather than refraction to produce an image.)

Refraction of a light ray occurs when light traveling in one medium strikes the surface of a second medium. The amount of refraction is determined by the speed of light in the two media (*media* is the plural of *medium*) and the angle with which the ray strikes the surface. The speed of light is related to a quantity called the index of refraction. The index of refraction n is defined as the ratio of the speed of light in vacuum c to the speed of light in the medium, v, given by

$$n = \frac{c}{v}. \tag{2}$$

Mathematically, we see that the speed of light in a medium is inversely proportional to the index of refraction, so that the larger the index of refraction, the smaller the speed of light. Vacuum has an index of refraction of exactly 1, while water has an index of about 1.33, and diamond about 2.42. Air has an index that is nearly the same as vacuum, depending slightly on the density of air, as well as pressure and temperature. You may have noticed a shimmering effect when looking across a parking lot or over the surface of a car on a hot summer day. This shimmer is a result of the refraction of light through different parts of the air with slightly different values of the index of refraction.

We can use equation (2) to determine the speed of light in various media, rearranging it as $v = c/n$. Because the speed of light in vacuum is 3×10^8 m/s, the speed of light in water is $3 \times 10^8/1.33 = 2.26 \times 10^8$ m/s, and in diamond it is 1.24×10^8 m/s.

A light ray will be refracted when it travels in one medium at a certain speed and then strikes the surface of a second medium where it slows down or speeds up. The greater the change in speed, the greater will be the change in direction of the ray. Different colors of light have slightly different indices of refraction in a medium, which causes them to refract at slightly different angles. This small difference based on color is why we see the various colors displayed in a rainbow, or through a prism, and one of the reasons why diamonds are so beautiful.

The amount of refraction depends not only on the change in the speed of light, but also on the angle at which the ray strikes the second medium. When light is traveling perpendicular to the surface, it does not refract, but continues moving in the same direction. If the light hits the surface at an angle as measured from the perpendicular direction, it will refract. The perpendicular direction has a special name, called the *normal*; the word "normal" is a mathematical term that means "perpendicular." It turns out that for larger angles at the surface as measured from the perpendicular direction, the refraction is greater. The largest refraction occurs when the light ray is moving nearly parallel to the surface at an angle of 90° as measured from the normal.

One of the laws of optics that spells out the relationship of the speed and the angle of the light moving in two media is given by *Snell's law*, also called the *law of refraction*. The law can be written

$$n_1 \sin\theta_1 = n_2 \sin\theta_2, \tag{3}$$

where n_1 is the index of refraction of the first medium, n_2 is the index of the second medium, θ_1 is the angle measured from the normal in the first medium (called the *angle of incidence*), and θ_2 is the angle that the refracted ray makes measured from the normal in the second medium (called the *angle of refraction*) (fig. 6.2).

What about some examples? When light is traveling in air and strikes the surface of water, the light slows down because the index of refraction increases; the ray bends so that the angle of refraction is less than the angle of incidence. When light is traveling from water into air, the angle

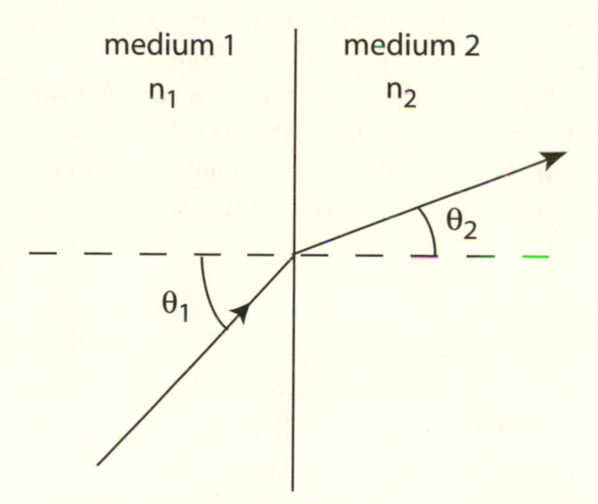

FIGURE 6.2. Refraction of light at the boundary (represented by the vertical solid line) separating two transparent media. A ray of light traveling in medium 1, which has an index of refraction n_1, strikes the surface of medium 2, which has an index of refraction n_2, and bends or refracts as it enters medium 2. The incident ray of light traveling in medium 1 makes an angle of θ_1 measured from the normal (represented by the horizontal dashed line). The refracted ray in medium 2 makes an angle of θ_2 measured from the normal.

increases. When light is traveling perpendicular to the surface so that the angle of incidence is zero, it does not refract, but continues moving in the same direction. When we look outside a window, the light that enters our eyes has been refracted as it passes through the window. On the outer surface it bends by a small amount and when it emerges it bends the opposite way by a small amount as well. If the two window surfaces are parallel to each other, then the direction of the emerging ray is parallel to the direction of the initial ray, but it is offset slightly (fig. 6.3a). When the two surfaces are not parallel, as in the prism shown in figure 6.3b, the emerging ray moves in a different direction. It is this changing direction of the emerging ray that allows an optical device like a lens to refract light with certain characteristics, based on symmetry and the angle, in order to form an image of an object.

So what does all this have to do with the eye? When we look at an object, light comes from it (either emitted by it as by a light bulb or a candle, or reflected from it as from a book or a person) and enters the eye. A light ray has to go through several different transparent media before and after it strikes the eye, and at each interface refraction takes place. Most of the refraction occurs when light goes from air into the eye because that is where the greatest change in index of refraction takes place. When light strikes the other surfaces inside the eyeball only a small amount of refraction occurs. However, this small amount of additional refraction, especially in the lens, is necessary to form an image.

Refraction by a Lens

Because light entering the eye first encounters the cornea, which is a curved surface, and later passes through the lens of the eye, it helps to understand how a lens changes the direction of a ray of light, and how it can form an image. Refraction of light by various types of optical devices, such as a magnifying glass, a camera, a telescope, the eye, and eyeglasses, involves at least one lens. For most of our discussion, we will consider spherical lenses, which really aren't spherical themselves, but have surfaces that are portions of larger spheres (fig. 6.4).

The two surfaces of a lens can be convex (outward curvature), concave (inward curvature), or flat (no curvature). Spherical lenses can take

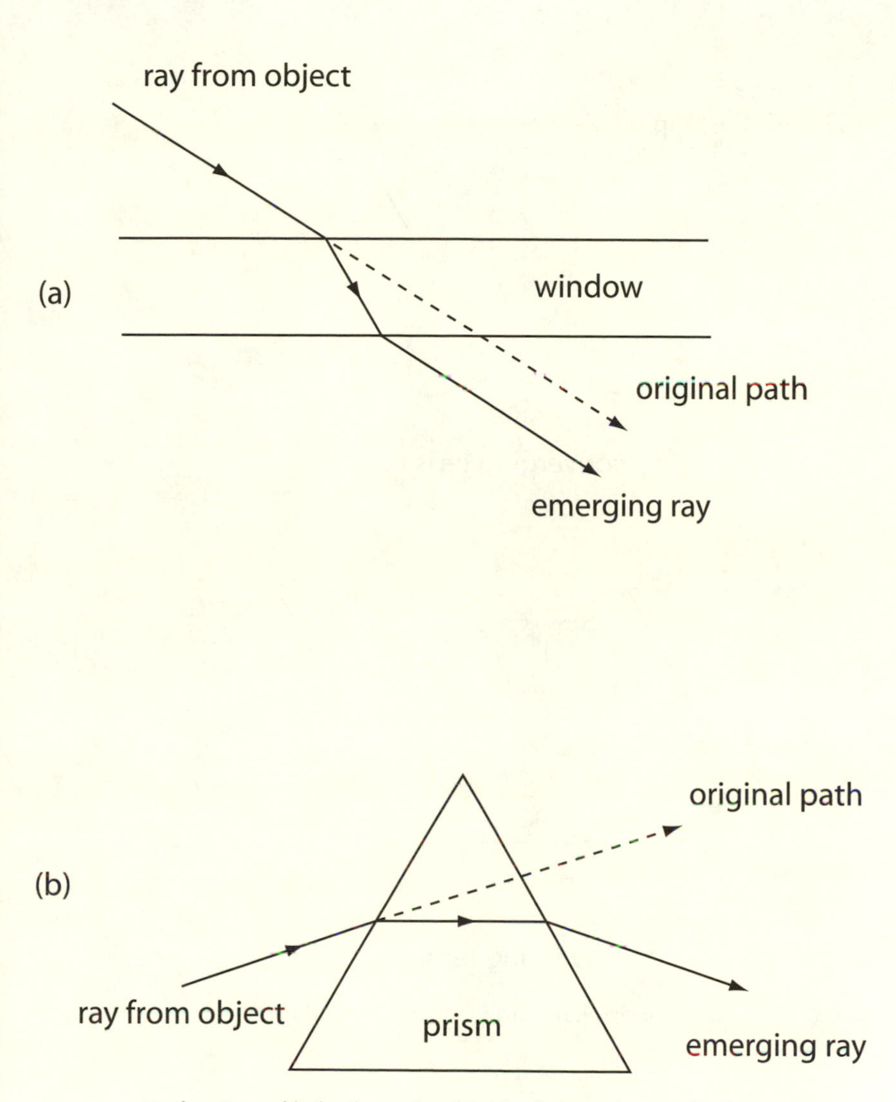

FIGURE 6.3. Refraction of light through a block of glass. A ray of light coming from an object refracts through a block of glass that has parallel surfaces, such as a window. The emerging ray is parallel to the original path of the ray. For a light ray that passes through a block of glass with nonparallel surfaces, the emerging ray moves in a very different direction than the original ray.

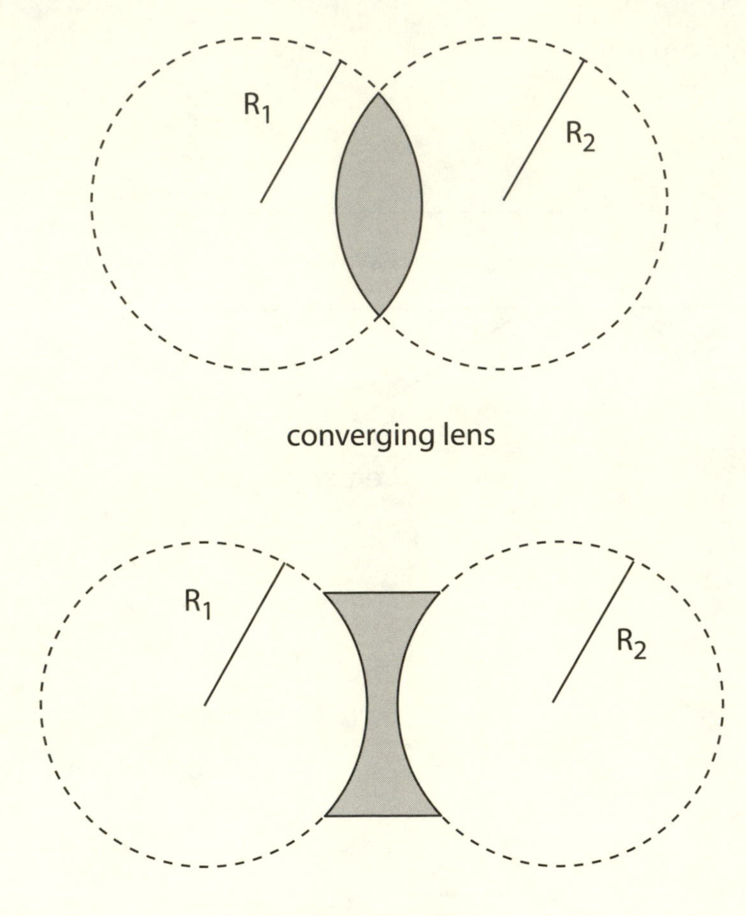

FIGURE 6.4. Spherical lenses have surfaces that are portions of spheres with radii R_1 and R_2.

on several different shapes: (1) biconvex, where both surfaces are convex, (2) biconcave, where both surfaces are concave, (3) plano-convex, where one surface is flat and the other convex, (4) plano-concave, where one surface is flat and the other concave, (5) convex meniscus, where one surface is convex and the other concave and the middle of the lens is thicker than its edges, and (6) concave meniscus, where the middle of the lens is thinner than its edges (fig. 6.5).

Lenses can be divided into two groups: converging lenses and diverging lenses. A converging lens causes light from an object located very far away to converge to a point near the lens, called the focal point. If you have used a magnifying glass to focus light from the sun so that you could burn leaves or paper, you have put this principle into action. These focused rays of light form a real image of the sun, and contain enough energy to cause damage! A diverging lens causes light rays from a distant object to diverge, or spread out from their original direction. Sunlight that passes through a diverging lens forms a halo-type display on a piece

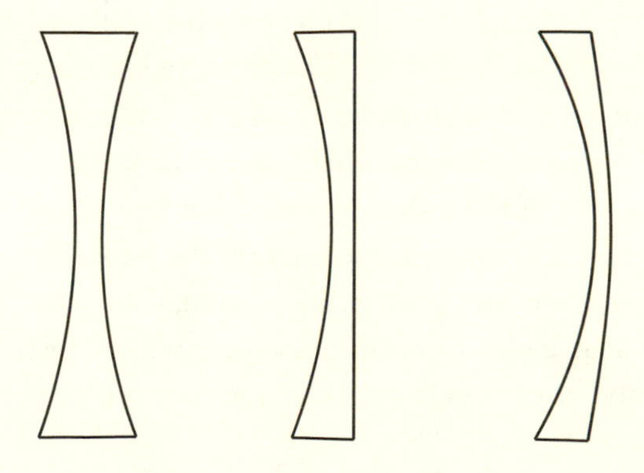

FIGURE 6.5. Common shapes of lenses. Converging lenses are thicker in the middle than on the edges, while diverging lenses are thicker on the edges compared to the middle.

of paper. I can remember being in grade school and playing outside on a sunny day with sunlight passing through my glasses, which were diverging lenses, and seeing the halo effect around the shadow of my head as I looked at it on the ground.

The rule to determine whether a lens is converging or diverging is this: if it is thicker in the middle than on the edges, it is converging; and if it is thinner in the middle than on the edges, it is diverging. With this rule in mind, we see that biconvex, plano-convex, and convex meniscus lenses are converging lens; biconcave, plano-concave, and concave meniscus lenses are diverging lens. That explains why a magnifying glass is a converging lens (biconvex) and my eyeglass lenses, being diverging lenses, are so thick around the edges.

Again, what does this have to do with the eye? The eye acts as a converging lens, causing the light rays to converge to form an image on the retina at the back of the eyeball. Some people have to wear eyeglasses that are converging lenses in order to see clearly, while others have to use diverging lenses. We'll look at this idea in more detail later.

Real and Virtual Images

For a person to see clearly, an image must be formed on the retina of the eye. If a person cannot see clearly, then he or she often has to wear eyeglasses, which form their own image. The person looks through the eyeglasses at this image and the eye forms an image on the retina. The image formed by the eyeglasses is called a *virtual image* and the image formed on the retina is called a *real image*. Both real and virtual images have certain characteristics that help to identify them. A real image occurs when light from an object passes through a lens and converges to form the image on the opposite side of the lens from the object. If a piece of paper is held behind the lens, the image can be seen on the paper. A virtual image occurs when light from an object passes through the lens and diverges (spreads out) as if it came from an image located on the same side as the object. The only way to see a virtual image is by looking through the lens.

Converging lenses form real images. When rays from the sun pass through a magnifying glass they converge to a small point that is a real image of the sun. A camera uses a converging lens to form a real image on the film. A slide projector, overhead projector, or data projector forms a real image on a screen for all in a room to view. Our eyes form real images on the retina of the eye. The real image formed by a single converging lens is upside down, or inverted, compared to the original orientation of the object (fig. 6.6a). The image formed on the retina of the eye is also upside down.

But a converging lens can form a virtual image also. When the lens is held fairly close to an object, a virtual image is formed. If you look through a magnifying glass at the small print in a book, the larger letters that you see are virtual images; they can only be seen by looking through the lens. A virtual image formed by a converging lens is upright with the same orientation as the object and is larger than the object (fig. 6.6b).

A diverging lens forms only virtual images. These images are smaller than the objects and are upright compared to the direction of the object. If you look through a diverging lens, everything you see is smaller than the actual object (fig. 6.6c).

Some optical devices use more than one lens. The combination of lenses can be used to produce images that have special characteristics. For example, microscopes use at least two lenses to produce large images so that very small objects can be seen. Telescopes use combinations of lenses to form a virtual image of a very distant object. The large lens of a nice camera actually consists of several lenses encased together in order to give a real image on the film or detector array.

These concepts are important for understanding how images are formed by the eye. Simply put, the eye acts as a converging lens that forms a real, inverted image on the retina. If this image is not focused properly, then eyeglasses or contacts can be used to form a virtual image in front of the eyeball; the eye can now observe this image and then produce a focused real image on the retina.

Focal Length of a Lens

The ability to form a clear image on the retina of the eye also depends on the focusing properties of the eyeball. The focal length and the power of a lens are two parameters that are used to describe the focusing ability. The

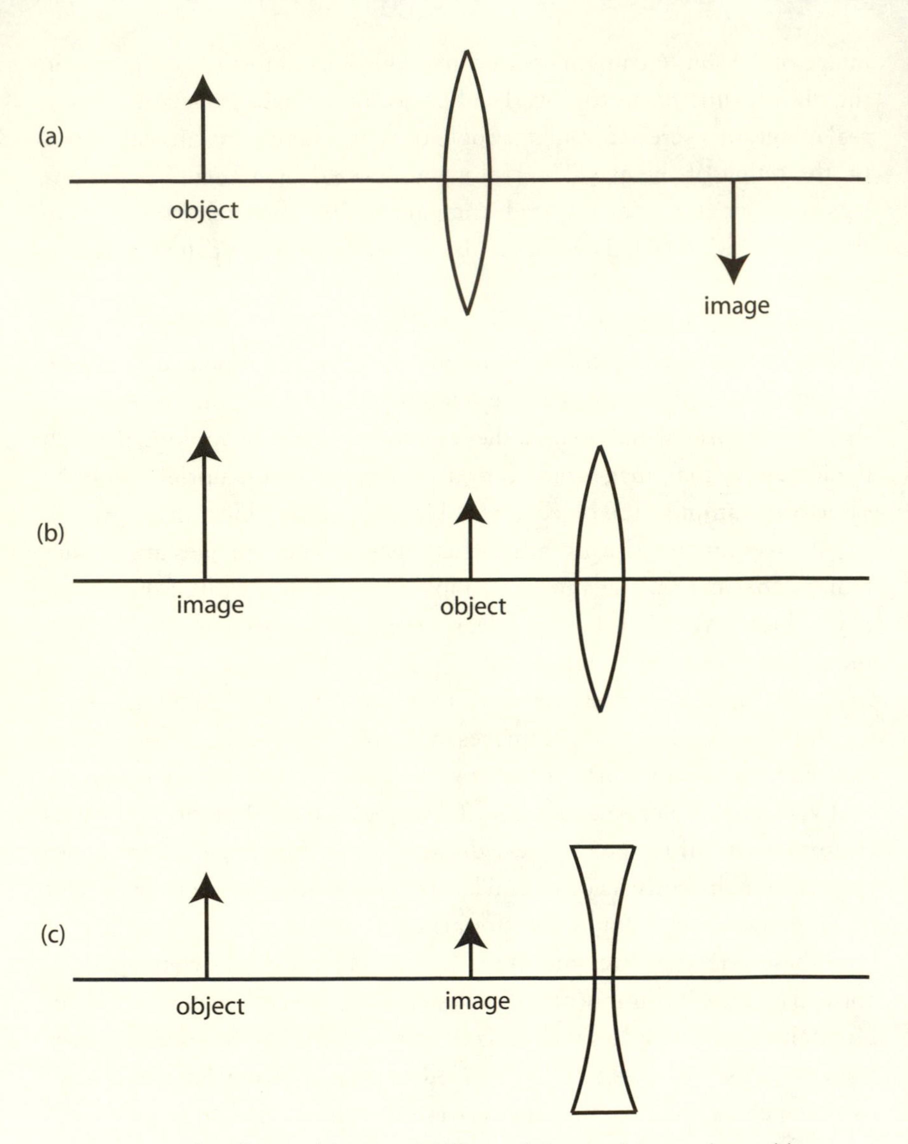

FIGURE 6.6. Real and virtual images. (a) The real, inverted image formed by a converging lens is located on the opposite side of the lens from the object. (b) The virtual, upright image formed by a converging lens is located on the same side as the object and is enlarged. (c) The virtual, upright image formed by a diverging lens is smaller than the object.

focal length of a lens is related to how curved the surface is. Let's look at these properties to see how they relate to the eye.

When light from a very distant object strikes a lens, the rays coming from the source are nearly parallel to each other. When they emerge from a converging lens, they converge to a point, which is called the *focal point*. The distance from the center of the lens to the focal point is called the *focal length*. The example of light from the sun passing through a magnifying glass applies here; the image of the sun is very small compared to the size of the actual sun itself, and the distance to the sun is huge (about 93 million miles) whereas the distance to the image is short, although not zero.

Lenses can be made to have a certain focal length, which depends on the curvature of the front and back surfaces of the lens. The mathematical relationship between these parameters is expressed in the lens maker's equation, given by

$$\frac{1}{f} = (n-1)\left(\frac{1}{R_1} - \frac{1}{R_2}\right), \qquad (4)$$

where f is the focal length, n is the index of refraction of the material used to make the lens, and R_1 and R_2 are the radii of curvature of the front and back surfaces of the lens. So, to make a lens with a certain focal length, the lens maker chooses a certain type of transparent material, and then grinds and polishes each surface so that it has a specific radius of curvature.

Consider a biconvex lens where the front and back surfaces have the same radius of curvature. There are some special rules dealing with positive and negative signs, specifically, set $R_1 = -R_2 = R$, so that the focal length in equation (4) can be written as

$$f = \frac{R}{2(n-1)}. \qquad (5)$$

The main point that I want you to see in this equation is that the focal length of a biconvex lens is proportional to the radius of curvature of the surface of the lens. This mathematical relationship indicates that the flatter the lens, that is, the larger the radius of curvature, then the longer the focal length; and the more highly curved or fatter the lens, that is, the smaller the radius of curvature, then the shorter the focal length.

Another quantity related to the focal length of a lens is called the power P of the lens. It is defined mathematically as

$$P = \frac{1}{f}. \tag{6}$$

Power can be thought of as the "strength" of a lens. For someone who wears glasses or contacts, it is the "prescription" given by an optometrist or discussed by an optical technician. The power is stated as a number on a box of contact lenses, such as –4.5 D for someone who is quite near-sighted. We will discuss the units later.

Because power and focal length are inversely proportional, we see that a lens with a short focal length has a large power, and a lens with a long focal length has a small power. Don't confuse power with magnification, which has to do with the size of the image compared to the object. The idea here is that the more highly curved (fatter) the lens, that is, the smaller R, then the shorter the focal length and the greater the power. Conversely, for a lens with a smaller curvature (flatter), which means that R is longer, the focal length is long and the power is smaller.

So how does this apply to the eye? The lens of the eye is similar to a biconvex lens. Muscles in the eye can change the shape of the lens from being flatter to being fatter, or vice versa. Thus the focal length of the eye can change, which is very important for seeing objects that are close to the eye (when the lens is fatter), as well as for seeing objects that are far away (when the lens is flatter).

One other thing related to curvature that is very important concerns the shape of the cornea, the front surface of the eyeball. Most refraction occurs at the cornea (which has a convex shape): if the cornea is too highly curved then its focusing ability is too strong and the image forms in front of the retina; if the cornea is too flat then its focusing ability is too weak and the image forms behind the retina. In both cases, the light that strikes the retina is not in focus and the image appears blurry to the individual.

Thin Lens Equation

The ability of the eye to produce a focused image on the retina depends not only on the focal length of the eye, but also on how far the object is located from the eye and on the distance to the image location. These quantities are related. Suppose we have an object located at a certain distance from a spherical lens. How can we determine where the image will be formed by the lens? One way to do this is to use the *thin lens equation.* This equation relates the focal length of a lens, f, the distance from the object to the lens, d_o (called the object distance), and the distance from the image to the lens, d_i (called the image distance), in the following way:

$$\frac{1}{f} = \frac{1}{d_o} + \frac{1}{d_i}. \tag{7}$$

Because most lenses have fixed focal lengths (as determined by the lens maker), for objects at various distances from the lens, the image distances must change as well. That is the main reason why you need to adjust the "focus" of a telescope, or binoculars, or a microscope—you aren't really changing the focal length of the lenses used, but rather the distance from the lens to the image.

However, for the eye, it is the image distance that is fixed because the distance from the lens to the retina is constant (at least over the short time required to view a close object and then to view an object that is far away). Thus, to focus on an object the focal length of the lens of the eye must change depending on where the object is located. In order to change the focal length of the eye, the shape of the lens must change. To see objects that are close to the eye, the lens must be fatter, and the focal length of the lens is quite short; to see objects that are far away, the lens must be flatter, and the focal length is relatively long. Note that the longest that the focal length can be (for normal vision) is the distance from the lens to the retina. To understand how the shape of the lens can change we need to know something about how the eye is made.

Anatomy and Physiology of the Eye

The eye, or eyeball, is similar in shape to a hollow sphere that is about 22 to 23 mm in diameter in the adult human. The wall of the eye is composed of three layers, the fibrous layer, the vascular layer, and the sensory layer. The eye is filled with two types of fluid, the *vitreous humor* and the *aqueous humor*, both of which are clear so that light can pass through them. Also in the interior of the eye are the *lens*, which is flexible, and the *ciliary body*, which changes the shape of the lens. The lens and its focusing components divide the eye into two parts, the anterior segment (the smaller front part of the eyeball) and the posterior segment (the larger rear part of the eyeball).

The fibrous layer, which is the outside layer of the eyeball, consists of the *sclera* and the *cornea*. The sclera is the white part of the eyeball and covers most of the eye (the posterior segment). The cornea is clear and colorless; it bulges outward from the eyeball and covers the anterior segment.

The vascular layer (the *uvea*) is next and consists of the *choroid*, the *ciliary body*, and the *iris*. The choroid contains most of the blood vessels for the eye and is brown in color. The dark color is necessary so that it absorbs most of the light striking it to keep the light from reflecting around in the eyeball. It covers the posterior segment. In the anterior segment is the ciliary body, which is just a continuation of the choroid, but contains the muscles and ligaments needed to change the shape of the lens. We will look at this part in more detail when we discuss the lens.

The *iris* lies between the lens and the cornea, and it gives the eye its color. It has a round opening called the *pupil* that allows light to pass through the lens; the pupil is open like the hole of a donut, although it is filled with fluid. The iris controls the amount of light that enters the eye. The smaller the iris, the less light can enter, and vice versa. The iris contracts and dilates automatically based on how much light enters the eye. The iris is important in the physics of image formation because, as it turns out, the larger the iris, the more distortion occurs. This distortion, called *spherical aberration*, occurs because the spherical surface of a lens does not produce the best in-focus image. Light rays that pass through a lens near its outer edges will refract more than rays that pass near the center of the lens. Thus, all the rays do not come to a proper focus in the eye and a distorted image is produced. In dim light situations, the iris is large to let more light into the eye, but the distortion makes objects blurrier, a bit out of focus. In bright light, the iris is small, there is less distortion, and objects appear sharper.

The innermost layer of the eye is the sensory layer that contains the *retina*. The retina covers most of the posterior segment of the eyeball and extends close to the ciliary body. The retina is the screen on which the image is formed. The outer part of the retina helps to absorb light like the choroid. The inner part of the retina (called the neural layer) consists of cells that absorb light and contain photoreceptors that convert light energy to electronic signals. There are two types of photoreceptors: rods and cones, whose names roughly describe their shape. The rods are very sensitive to light and are useful for dim light situations and peripheral vision; they are not used in determination of color. The cones are not as sensitive to light, so they operate best in bright light conditions; they are used to see finer details and are best at color detection. Light that strikes these photoreceptors produces an electric current that becomes a signal that is sent to the brain by the optic nerve.

There are three kinds of cones that are sensitive to different colors in the visible spectrum. Recall that the visible spectrum spans the range of wavelengths from 400 to 700 nm, with 700 nm being the edge of the red end of the spectrum and 400 nm being the edge of the violet end. The "blue" cones are most sensitive in the blue region at 440 to 450 nm and the "green" cones are most sensitive in the green region at 530 to 540 nm. The "red" cones are most sensitive in the yellow-green region at 560 to 570 nm, but are more sensitive to red light than the other cones. Light from objects of different colors strikes the retina, stimulating different numbers and kinds of cones. The mixture of these stimulations in the photoreceptors produces a signal that the brain interprets as color.

Also found in the retina are two regions called the *fovea* and the *optic disk*. The fovea is located at the very back of the eye, directly behind the lens, and is actually a very small point less than half a millimeter in size contained in a larger oval region called the macula. Here the sharpest vision occurs so that we can read, discriminate between colors, and see fine detail. The fovea has a high concentration of cones. The optic disk is located at the point where the optic nerve is attached to the retina. The optic disk is also called the blind spot because it does not have any photoreceptors. It does not cause a problem with sight because the brain fills in the image in that region even without any electronic signal there.

The vitreous humor and aqueous humor are fluids that fill the two segments of the eye. The front of the eye (the anterior segment) between the lens and the cornea holds the aqueous humor. The aqueous humor, as it name implies, is composed mostly of water and is a clear liquid similar to blood plasma. It provides nutrients for the tissues in the anterior segment, carries away waste from the area, and helps to maintain proper pressure in the eye (intraocular pressure). If the aqueous humor does not drain properly, it can cause excessive pressure in the eye that may cause damage to the retina and the optic nerve. This condition is called glaucoma and may cause blindness.

The vitreous humor is found in the posterior segment of the eye and fills the space between the lens and the retina. It is a gel-like fluid that contains a large amount of water. It does not flow like the aqueous humor, so it remains in the posterior segment. It is clear so that light moves through it and it helps to keep the retina in place by contributing to the intraocular pressure.

The Lens

The lens of the eye is located directly behind the iris (and pupil) between the anterior and posterior segments. It is transparent so that light can pass through it; it is flexible to allow the person to see objects that are close by or far away (this property is called *accommodation*); and it is biconvex (a converging lens) so that it can form a real image on the retina. The lens mostly consists of lens fibers, which extend from the front of the lens toward the rear and are packed in layers similar to the layers of an onion.

The shape of the lens is controlled by the ciliary body, which contains the ciliary muscle and ligaments (called ciliary zonules). The ciliary muscle is a muscular ring that circles the outside of the lens; it is attached to the lens by ligaments that form a halolike structure around the lens. When the ciliary muscle is relaxed, it has it largest diameter, so the ligaments are under tension, pulling outward on the lens and forcing it into its flattest shape. This is the situation that occurs when viewing distant objects. When the ciliary muscle is tense, it contracts and has its smallest diameter. This causes the ligaments to lose their tension, which results in less force on the lens and the lens relaxes to its natural shape, which is fatter. This is necessary in order to focus on objects that are close to the eye.

As a person ages, new lens fibers are added to the lens that cause it to become slightly larger, more compact, and denser. The lens loses some of its flexibility and it becomes difficult for the person to focus on objects that are close by. Also, the lens may become cloudy as the person ages. The cloudiness of the lens is called a *cataract*.

Path of Light into the Eye

Now that we have an idea about the structure and function of the eye, let's see what has to happen for a focused image to be formed on the retina. Light from an object must enter the eye and pass through five different transparent media before it strikes the retina: first, it leaves the object and moves through the air, it enters the eye through the cornea, passes into the aqueous humor, through the lens, and finally through the vitreous humor and on to the retina (fig. 6.7).

The indices of refraction of each material are as follows: for air, n = 1.00; for the cornea, n = 1.38; for the aqueous humor, n = 1.33; for the lens, n = 1.40; and for the vitreous humor, n = 1.34. There are several things to note about the indices of refraction of the different materials. First, the largest change in index of refraction occurs when light enters the eye at the cornea from the air, which means that most of the refraction occurs at that surface. The indices of refraction of the other media do not vary by much, so there should be only small (but very important) amounts of refraction at the other surfaces. The index of refraction of the

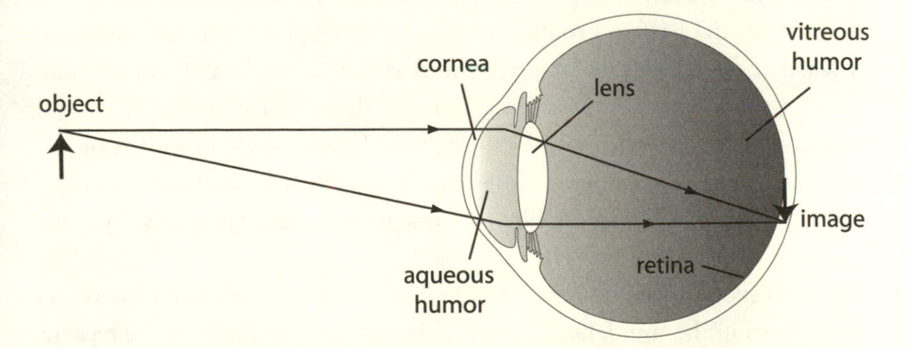

FIGURE 6.7. Cross-sectional view of the eye. The index of refraction of the cornea is about 1.38, of the aqueous humor is 1.33, of the lens is 1.40, and of the vitreous humor is 1.34.

aqueous humor (1.33) is identical to that of water, indicating the high content of water in that portion of the eye, as well as in the vitreous humor, which has an index of 1.34. The index of refraction of the lens (1.40) indicates that there is enough difference in indices of refraction with the surrounding materials (the aqueous humor and the vitreous humor) to cause enough refraction for focusing the light rays onto the retina.

Comparison of the Eye with a Camera

It is interesting to compare the eye with a single-lens camera. When these optical devices are operating correctly, entering light is focused onto a screen, forming a real image. The screen in the camera is the film or an array of detectors, whereas in the eye it is the retina. Let's look again at the thin lens equation, which relates the focal length of a lens f, to the object distance d_o, and the image distance d_i:

$$\frac{1}{f} = \frac{1}{d_o} + \frac{1}{d_i}. \tag{8}$$

Both the camera and the eye take light from objects that are at different locations, so that the object distance d_o varies. One of the other two parameters, focal length f or image distance d_i, must vary in order to satisfy the equation.

The camera lens has a fixed focal length, so in the thin lens equation f is constant. In order to produce a focused image, the lens must be moved toward or away from the object, as well as toward or away from the film, causing the object distance and the image distance to change. The final position of the lens with respect to the film depends on how far the object is located from the camera. Note that if the object is very far away the distance from the lens to the film (the image distance d_i) is nearly the same as the focal length of the lens.

However, for the human eye, the distance from the lens to the retina is fixed; so in the thin lens equation d_i is constant. In order to see clearly, the shape of the lens changes so that a well-focused image is formed directly on the retina. Thus, as the object distance changes, the focal length of the lens must change in order to maintain a constant image distance. For an object that is far away, that is for d_o large, the mathematics of the thin lens equation says that the focal length is nearly the same as the image distance, which is the longest focal length that occurs when the lens is flatter.

Near Point and Far Point

In order for a person to see clearly, the object must be located at a position such that the eye can form a focused image on the retina. If the object is close to the eye, the shape of the lens must become fatter in order to form a clear image. If the object is far away, the lens must flatten to produce a proper image. The changing shape of the lens is called accommodation.

Have you ever held a book so close to your eyes that you are not able to read it? Have you ever had difficulty seeing objects that are very far away? There is a range of distances where objects can be located so that we can see them clearly. An object must be located somewhere between two distinct points or distances for the eye to produce a well-focused image. These two points or distances are known as the *far point*, which is the farthest position from the eye, and the *near point*, the closest position to the eye. Objects located farther away from the eye than the far point will appear blurry to the viewer and so will objects placed closer to the eye than the near point.

To form an image of an object located at the far point, the surface of the lens must be in its least curved, or flattest, shape. To form an image of an object located at the near point, the lens must be in its most curved, or fattest, shape. If an object is located outside this range, the person will not be able to see it clearly; there is not enough accommodation by the lens of the eye. For a person with normal vision, the far point is very far away (i.e., infinity), and that person should be able to see distant mountain ranges or the moon and stars at night. The near point ranges from about 10 cm to 100 cm, depending on one's age.

Vision Problems

It is very important to be able to see clearly. But many people have problems seeing distant objects and others have difficulty seeing objects that are close, as when reading a book. Others see distortions when looking at an object from different directions.

Vision problems have been known for a long time. Early writings describe people who have problems seeing, particularly the elderly. The ancient Greeks and Romans, as well as biblical manuscripts, mention the elderly as having "dim eyes" and needing to write in large letters.

The use of a glass globe filled with water as a magnifier was reported in the first century; reports of magnifying glasses ("reading stones") began to appear around the year 1000 AD. In the thirteenth century, scientist and philosopher Roger Bacon mentioned that reading lenses are "useful to all persons and to those with weak eyes" for reading. Eyeglasses for seeing objects at a distance began to be used in the 1400s. Benjamin Franklin invented the bifocal lens in 1784 to avoid switching back and forth between his reading glasses and his distance glasses. Contact lenses were first described by Leonardo da Vinci in 1508, although they made their appearance in the 1800s. They were very uncomfortable and very unpopular for many years, but improved greatly when molds began to be made from living eyes in 1929. Plastic contacts were made in 1948 and soft contact lenses were commercially available in 1971.[1,2,3,4] Let's look more closely at several types of vision problem and various methods used to correct them.

Nearsightedness or Myopia

An individual who cannot see distant objects has a condition known as *nearsightedness*, or *myopia*. Refraction of light in the nearsighted eye causes the image to form in front of the retina, so that by the time the light reaches the retina, it is out of focus. This condition usually occurs when the shape of the eyeball is too long, or oblong, somewhat like a football. Another reason is that the cornea is too highly curved.

For a normal eye, the far point is at infinity, but not for the myopic eye. A person who is only slightly nearsighted may have a far point of, say, a few meters and typically will not require much correction. However, a person who is *very* nearsighted may have a far point that is less than 20 cm and likely would have difficulty reading a book held only 30 cm away. Certainly, this person would have trouble driving. I can remember when I was in third grade that I was not able to see the board clearly, so I had to sit in the front of the room. After getting glasses, I was amazed

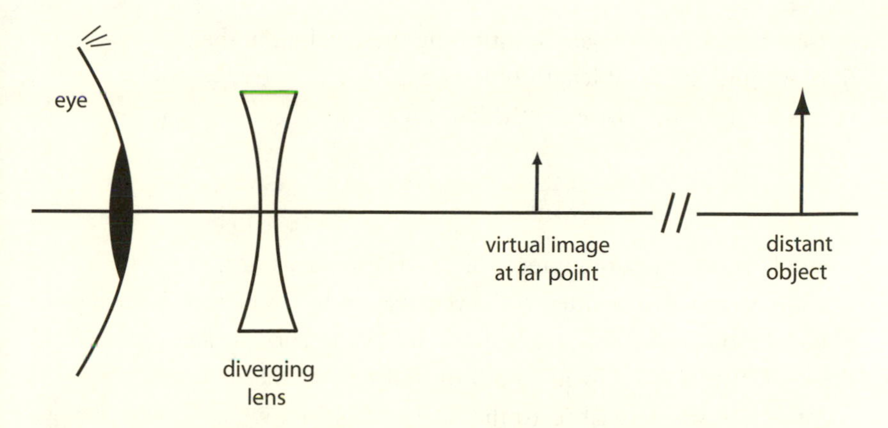

FIGURE 6.8. Correction of the nearsighted eye by use of a diverging lens. The object is located very far away, but a virtual image is formed near the far point. The eye can now look at this image and form a real image on the retina.

at all the things I could see, particularly fine details like power lines hanging from pole to pole and mortar joints on brick houses!

To correct for myopia using contact lenses or eyeglasses, a diverging lens is placed in front of the eye. This lens causes incoming rays to spread out from their original direction, as if they came from a virtual image in front of the lens. If the focal length of the diverging lens is chosen properly, the virtual image is formed near the far point of the eye. The virtual image of the diverging contact lens or eyeglass becomes the object for the next lens, which is the eyeball itself. The eye can now take light from the virtual image located at the far point and form an image on the retina that is in focus (fig. 6.8).

Suppose a person is nearsighted with a far point of 30 cm. We want to determine the focal length of a lens that will allow this person to see distant objects clearly. The numbers turn out to be slightly different if the person is being fitted for contact lenses or for eyeglasses.

Consider contact lenses first. To find the power of the lens needed for proper correction, we start with the thin lens equation

$$\frac{1}{f} = \frac{1}{d_o} + \frac{1}{d_i}.$$

With the object located very far away, we set $d_o = \infty$. The virtual image needs to be formed at the far point, so we set $d_i = -30$ cm (negative because it is a virtual image). Substituting these values in the thin lens equation, we find the focal length of the contact lens to be $f = -30$ cm, which corresponds to the power of the lens given by

$$P = \frac{1}{f} = \frac{1}{-30\,\text{cm}} = -\frac{1}{3\,\text{m}} = -3.33\ \text{D},$$

where D is the unit of diopter, which is the same as m^{-1}.

If a person is to be fitted with eyeglasses, which sit about 2 cm in front of the eye, the image needs to be at about 28 cm from the lens (30 cm from the eye). We find the focal length of −28 cm, or the power of −3.57 D. Most lenses are prescribed to the nearest 0.25 D, which is why the eye doctor (optometrist or ophthalmologist) often switches back and forth between a couple of lenses and prompts the patient about which lens helps them to see more clearly.

Soft contact lenses are dispensed in boxes labeled with the power of the lens in diopters. The negative sign for the focal length and for the power indicates that diverging lenses are needed to correct nearsightedness.

It is fairly easy to tell if a person who is wearing glasses is nearsighted by recalling several features about diverging lenses. The edges of the person's glasses should be thicker than the middle of the glasses. If you cannot get close enough to tell, recall that when looking through a diverging lens, the image is smaller. If you look at the person's eyes, they should appear smaller than normal. If you still cannot quite tell, try to look through the lens at the side of the face. If it is easy to see the side of the face (the image is smaller than the object), the person is wearing diverging lenses.

Farsightedness, or Hyperopia

People who have trouble seeing objects that are close have a condition known as *farsightedness*, or *hyperopia* (also known as hypermetropia). In this situation the light is not refracted enough, and the rays converge to form an image behind the eye; the retina blocks these rays before they completely converge, causing the rays to be out of focus. This condition occurs if the eyeball is too short, if the cornea is too flat, or if the lens is not able to be made fat enough.

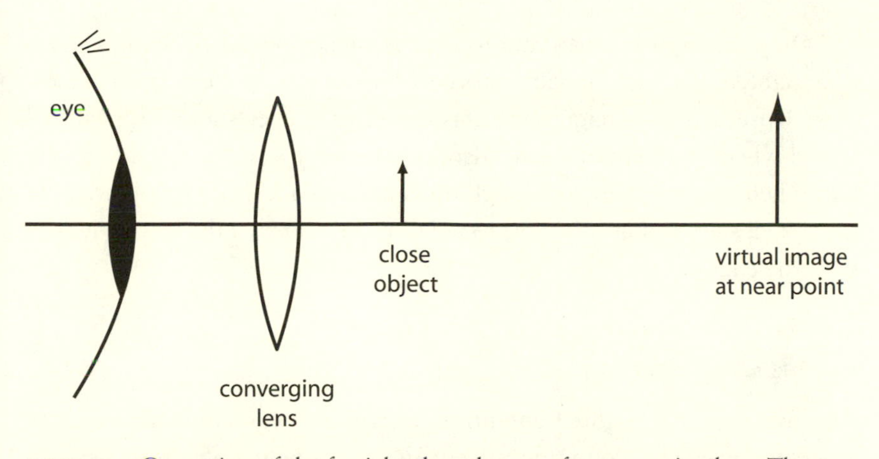

FIGURE 6.9. Correction of the farsighted eye by use of a converging lens. The object is located quite close to the eye, but a virtual image is formed at the near point. The eye can now look at this image and form a real image on the retina.

The farsighted individual has a near point that is farther away than normal. For example, a farsighted person whose near point is 80 cm will have great difficulty reading a book held at a comfortable distance about 30 cm away.

Hyperopia is corrected with a converging lens. Light from a real object diverges as it approaches the eye, but the converging lens causes these rays to diverge less. These rays appear to come from a virtual object located farther from the eye than the real object. If this virtual object is at the near point, the eye can form a clear image on the retina (fig. 6.9).

Suppose that a certain individual has a near point of 80 cm, indicating hyperopia. In order for the person to see an object held at 20 cm clearly, the focal length of a lens that can produce a virtual image at the near point needs to be calculated. In the thin lens equation, we use d_o = 20 cm and d_i = −80 cm for contact lenses. These values give us a focal length of f = +26.7 cm, or a power of P = +3.75 D. If eyeglasses are used instead of contacts, and if we assume that the glasses sit about 2 cm in front of the lens, then with d_o = 18 cm and d_i = −78 cm, we find a value for the focal length f = +23.4 cm, which gives a power of P = +4.27 D. The positive sign confirms that a farsighted person requires positive or converging lenses.

Once again, it is fairly easy to determine if a person wearing glasses is farsighted. Because farsightedness is corrected using converging lenses, these lenses act like magnifying glasses, and the eyes should appear enlarged when you look at them. Also, it is very hard to see the side of the face when you try to look through the lenses. In addition, if you are able to look closely, you may be able to tell that the center of the lens is thicker than the edges.

Age-related Problems

Presbyopia is a farsighted condition where the near point increases as a person ages. This condition affects almost all people, from those who have had normal vision (for their first 40 to 50 years) to those who are very nearsighted. It results from a lack of flexibility of the lens in the eye. Recall that the ability to see objects up close requires the lens to become more rounded, or fatter. The muscles controlling the shape of the eye are in their most tense orientation when you look at close objects, but that means that the ligaments are under no tension; the unstressed shape of the lens is the fatter shape. To look at objects that are far away, the muscles are relaxed and bigger around, which causes the ligaments to be under tension; they exert forces on the lens causing it to be stretched out, making it flatter. As a person ages the lens becomes more and more rigid with the lens remaining in a flatter configuration and not able to change to a fatter shape. Thus the person is not able to see objects up close as clearly.

To correct for presbyopia, a person may need reading glasses, which are converging lenses, just as a person with hyperopia requires converging lenses. You could say that presbyopia is a special kind of farsightedness that can be corrected in a similar way.

There are situations, however, where an aging individual who has difficulty seeing distant objects and who already wears glasses to correct for this myopic condition may develop presbyopia and have difficulty seeing objects up close. Bifocals are used in this case; the lower portion of the lens is used for close viewing and the upper portion for distance. Even trifocals are worn sometimes for people to see intermediate distances, as when working at a computer.

Astigmatism

Another common problem is *astigmatism*, which causes a distortion of the image along a particular axis. This problem occurs when the cornea or the lens is not spherical in shape, but somewhat oval. There are different focal lengths for different axes, which are often at right angles to each other. When an astigmatic person looks at a set of crossed lines, certain lines appear fuzzier than others and, if the picture is rotated, other lines will become fuzzy and the original ones will become clearer.

To correct for astigmatism, glasses or contacts that are not spherical lenses are used. In fact, they are often called cylindrical lenses, although they are not actually cylindrical in shape. These lenses have a different radius of curvature along two different axes that allow correction of vision for the different focal lengths of the eye.

Other Methods of Vision Correction

Because most refraction of light entering the eye occurs at the cornea, several vision corrective techniques have arisen that involve reshaping the cornea to cause more or less refraction. One such popular procedure is LASIK, but a brief look at earlier methods may help us understand why LASIK is so popular these days.

One of the first methods used to reshape the cornea was radial keratotomy (RK). This procedure involves using a knife to make incisions on the surface of the cornea, pointing radially outward like spokes on the wheel of a bicycle. The method allows the cornea to flatten somewhat. Think of frying bologna—if you place a slice of bologna on a frying pan, it will curl up forming a cup, but if you cut the edges, the tension forces are relieved and the bologna flattens against the pan. RK was used to correct for nearsightedness because if the cornea flattens a bit the curvature decreases. This method is difficult because is relies heavily on the surgeon to make precise incisions. Problems after this procedure are common and include infection, glare (especially at night), double vision, and scarring.

Another method that came along later is called PRK, or photorefractive keratectomy. This procedure used a laser to blast away tissue from the surface of the cornea in order to reshape it. PRK could correct for myopia as well as astigmatism. Early attempts at PRK resulted in clouded vision and scarring, but it achieved better results than RK. More recently, a similar procedure called LASEK (laser epithelial keratomileusis) has been developed and does not cause as many problems as PRK.

LASIK, laser-assisted intrastromal keratomileusis, is now the standard of vision correction by reshaping the cornea. The procedure uses a knife blade called a microkeratome, which is similar to a carpenter's plane, to shave a thin layer of the cornea so that it can be folded back out of the way. The laser beam blasts away at the subcorneal tissue, reshaping the cornea in such a way that when the flap is folded back into place, the resulting surface has the proper shape. This procedure can be used to correct for nearsightedness, astigmatism, and farsightedness. Not as many problems occur from this procedure because the flap seals easily and heals quickly, keeping infections to a minimum. The flap is large enough that regions where the cut occurred are out of the field of view of the eye (fig. 6.10).

LASIK does not prevent the development of presbyopia, so if this procedure is done in early adulthood, there will still be a need for reading glasses. However, LASIK can be performed so that one eye is corrected to allow for close vision and the other for far vision. This idea may seem a bit strange, but different corrections for each eye are already in use for people who wear contacts.

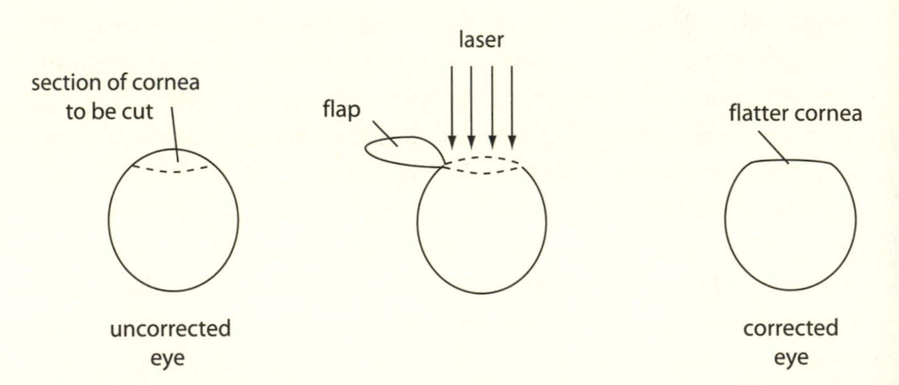

FIGURE 6.10. The LASIK procedure. It involves cutting a flap in the surface of the cornea, folding the flap back, and blasting away corneal tissue below the surface with a laser. The flap is folded back into place to give the correct curvature of the cornea.

Many other techniques are available to correct vision by changing the eye in some way. A common procedure is lens implant, where the lens of the eye is replaced with the proper curvature. (Lens implants are most often used for correcting problems from cataracts.) Another technique is orthokeratology where a contact lens is worn for several hours to flatten the cornea (to treat nearsightedness) so that vision is clear for an extended time; however, it is temporary. Conductive keratoplasty (CK) uses radio-frequency electromagnetic waves to make the cornea more curved (for farsightedness), but only on a temporary basis. Corneal implants can be used to flatten the cornea by placing circular plastic rings under the surface of the cornea, raising the outer edge, thus correcting for nearsightedness.

Other Problems

The vision problems just discussed can be corrected by making changes to the light entering the eye to produce an image that is in focus. However, there are many other types of problems that people experience. Most of these do not focus on refraction of light, but are more related to overuse, defects, disease, or injury.

To see objects that are up close several sets of muscles need to work, and this can lead to eye strain. These muscles act to control accommodation (which we have already discussed), the size of the iris or pupil, and convergence. Recall that the iris contracts and dilates according to the amount of light that enters the eye; but the larger the iris the more spherical aberration occurs. Light rays from an object that is close to the eye diverge from the object; as these rays enter the eye they continue to spread out so that many of them will likely go though the outer edge of the lens, thus causing significant spherical aberration. To help, muscles act to reduce the size of the iris so that light rays pass mostly through the center of the lens resulting in less distortion.

Convergence is the act of moving both eyes together to focus on an object up close. When the eyes are looking at an object at a distance, they are both looking in a direction that is almost perpendicular to the eye socket. But to look at an object that is very near the face, both eyeballs have to turn inward (if the object is in line with the nose), perhaps even becoming “cross-eyed.” Again, muscles that control the motion of the eyeballs act to provide proper convergence.

A *cataract* is a clouding of the lens that reduces its transparency. The cloudiness can range from minimal with only slight distortion of light to near opacity causing blindness. The clouding of the lens occurs if there are not sufficient nutrients supplied to the fibers of the lens. The fiber material clumps together into large enough particles that light is scattered off them, as off droplets of water in a cloud or fog. Cataracts have a number of different causes: ultraviolet exposure, diabetes, age, smoking; someone may be born with them. Cataract surgery is fairly common these days; the lens is removed and replaced with a plastic lens.

As mentioned previously, glaucoma is a condition where the optic nerve is damaged. It is usually caused by increased pressure in the eyeball, but it can occur in people with pressures in the normal range (12 to 22 mm-Hg). The primary method of treating glaucoma is to lower the intraocular pressure with medication (such as timolol). Surgical procedures can help to lower the pressure as well.

Macular degeneration is a condition that affects the macula of the eye. Recall that the macula is a region located in the retina directly behind the lens containing the fovea where the sharpest vision occurs. The degeneration of the macula involves thinning of the macular tissue, deposits of pigments, and/or the growth of unwanted blood vessels. This problem results in loss of central vision with no change in peripheral vision. Treatment is aimed primarily at prevention or slowing down the progression of the problem.

Other problems include inflammation of the eye from various types of bacteria. River blindness is caused by a worm in west and central Africa. Opacity of the cornea can also develop. A torn retina or a detached retina can occur if the retina thins or if the vitreous humor shrinks—recall that the vitreous humor helps to maintain pressure in the eye. Damage to the retina can also occur as a complication of diabetes or from a severe blow to the eye.

Another common problem is color blindness, which is not really blindness to color, but rather an inability to see all colors properly. The range of colors that a person can see depends on the amount of stimulation that each type of cone in the retina (red, green, and blue) receives. A person who is color blind has a deficiency or absence of one or more types of cone, or a shift in peak sensitivity. The most common form is red-green color blindness resulting from problems with the red and green cones.

This type of color blindness makes it difficult for a person to distinguish between various shades of red, yellow, and green. The individual can often compensate by recognizing shades of intensity. Color blindness is more common in men (about 5 to 10%) than in women (less than 1%).

Summary

Sight is another important sense of the body. Reading, watching a sunset, and seeing a movie are such enjoyable events. Hopefully, the concepts that I have described in this chapter will help you understand more about the optics of the eye. As with the other areas we have discussed, this topic can be explored in much more detail. Optometrists and ophthalmologists spend their entire careers studying the eye and caring for patients with normal vision and with vision problems. Individuals who have limited or no sight can often live independently and pursue life goals provided they have adequate training.

SEVEN

Biological Effects of Nuclear Radiation

Nuclear power, cobalt treatments, x-rays, irradiated food—exposure to radiation is something to be avoided at all costs, right? Well, not exactly. While exposure can cause problems, and we need to minimize the amount, radiation can be useful in providing energy, treating cancer, peering into the body, or killing unwanted bacteria.

Nuclear radiation is a term used to describe high-energy particles and photons, such as gamma rays, that are emitted by the nucleus of an atom. While x-rays do not involve the nucleus, they behave in ways that are similar to gamma rays. Often the term "ionizing radiation" is used to describe nuclear radiation and x-rays; therefore much of the discussion in this chapter will apply to x-rays as well as nuclear radiation. Ultraviolet (UV) rays and to some extent visible light can be considered ionizing radiation also, but their energy is much smaller than that of x-rays; the topics addressed in this chapter do not apply to UV rays or visible light.

In this chapter, we will look at some basic physics topics related to nuclear radiation, and see what effects it has on the human body. We will look at exposure rates, health effects, and medical applications including therapeutic and diagnostic techniques, such as treatment of cancer and methods of imaging.

The Nucleus

The atom consists of electrons, protons, and neutrons. Electrons orbit the nucleus of the atom, which is made up of protons and neutrons. There are two models of the atom to keep in mind while reading through this chapter: the planetary and the shell models. In the planetary model, we think of electrons orbiting the nucleus like planets orbiting the sun. This model arises from classical, or Newtonian, physics. In the shell model, we think of electrons existing in clouds or shells around the nucleus. This model arises from quantum physics, which includes ideas such as the uncertainty principle and probability. Electrons have a negative charge of $q = -e = -1.6 \times 10^{-19}$ coulombs, protons have a positive charge of $q = e = 1.6 \times 10^{-19}$ coulombs, and neutrons have no charge, $q = 0$. (The value of the charge e is sometimes called the fundamental unit of charge.) The charge of a neutral atom is zero because the atom has an equal number of negatively charged electrons and positively charged protons. Ions are atoms with extra electrons or that have lost electrons.

Electrons are involved in chemical and thermal processes and reactions. As atoms and molecules undergo chemical reactions, electrons are transferred between them. Many of the properties of the new chemicals formed are based on the behavior of the electrons in these new configurations and the bonds that are formed as atoms share or transfer electrons between them. Reactions may require heat to be added in order to proceed; other reactions may produce heat. Catalysts may be needed for certain reactions to occur. In all of these situations, electrons and their configurations within atoms and molecules are important.

The nucleus is composed of protons and neutrons, both of which are called nucleons (fig. 7.1). The number of protons determines the type of atom. For example, hydrogen has one proton, helium has two protons, carbon has six protons, iron has 26 protons, and so on. The charge of the nucleus is equal to the number of protons, Z, times the fundamental unit of charge e, or

$$q = +Ze.$$

Atoms with the same number of protons can have different numbers of neutrons; these atoms are called isotopes. We will discuss this topic in more detail later.

Energy of the Nucleus

The energies involved with the nucleus are very large compared to the energies involving the electrons. It does not take a lot of energy to remove

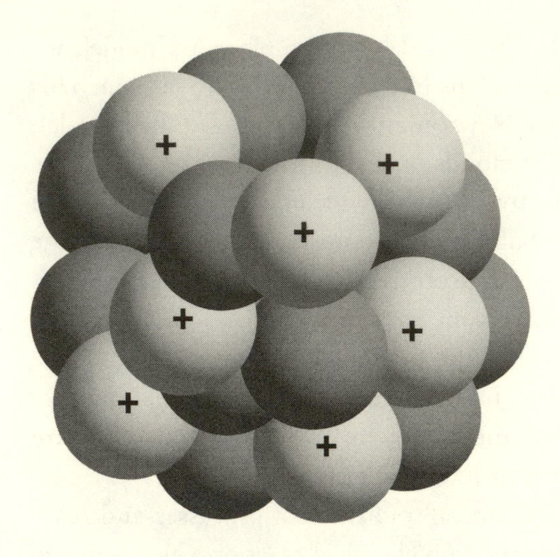

FIGURE 7.1. The nucleus of an atom is composed of neutrons (dark) and protons (light). Neutrons are uncharged; each proton has a charge of $+e$.

an electron from an atom, but much more to remove a proton or neutron. Typical energies for electrons are on the order of a few electron-volts, or eV, whereas energies involving the nucleus are in the range of millions to hundreds of millions of electron-volts. An *electron-volt* is the amount of energy an electron gains or loses when moved through an electric potential of one volt. For example, a 1.5 V battery provides 1.5 eV of electrical potential energy to a single electron. To remove a single electron from a hydrogen atom requires 13.6 eV of energy. By contrast the average energy needed to bind a proton or a neutron in the nucleus of a helium atom is about 7 MeV (seven million electron-volts). When uranium splits into two nuclei, more than 230 MeV of energy is released.

While energies in the nuclei of atoms are large when compared to electronic energies, they are relatively small when compared to the energies of larger objects. For example, the energy released by splitting a single uranium atom, 230 MeV, is about 4×10^{-11} J, while that of a baseball (mass of 145 g) moving at 85 mph is about 100 J. However, in a 100 g sample of uranium, there are about 2×10^{23} atoms of uranium. If all the energy of this sample is released at once, that is about 8×10^{12} J. By contrast, burning about 100 g of coal will give at most only a few million joules (10^6 J) of energy because that process involves energy from the electrons in the carbon, hydrogen, and oxygen atoms that make up the material.

One of the reasons that there is so much energy contained in the nucleus of an atom has to do with the size of the forces required to hold the nucleons together. Because protons are positively charged, they repel each other according to Coulomb's law and the law of charges, which states "like charges repel and unlike charges attract" (see chapter 5). In order for two or more protons to be held together in the nucleus, there must be a force that is stronger than the repulsive Coulomb force. This force is called the *strong nuclear force*. (There are four fundamental forces in nature—the first involving gravity, another involving electromagnetic forces, a third called the weak nuclear force, and last the strong nuclear force.) The strong nuclear force is a strongly attractive force between any two nucleons: between any pair of protons, between any pair of neutrons, and between any proton-neutron pair. This force is stronger than the Coulomb force (and the gravitational force) between any two nucleons, but only over very short distances of less than about 10^{-15} m. If the nucleus of an atom is larger than this size, it tends to be unstable and will emit nuclear radiation.

Because of the high energies associated with nuclear radiation, if it is absorbed by human tissue it can cause quite a bit of damage. The damage may result in harm to good tissue, resulting in cancer or other ailments. However, radiation can be used to treat cancer by destroying cancerous tissue; perhaps you know someone with cancer who has had cobalt treatments or other types of radiation treatment. We will discuss this topic in more detail later.

Nuclear Notation

Nuclear notation is a shorthand way of keeping track of important information about the nucleus. The notation takes the form

$$^{A}_{Z}X,$$

where X represents the chemical symbol, Z is the atomic number or number of protons, and A is the atomic mass number or number of nucleons (protons and neutrons). To get the number of neutrons N, we just subtract Z from A, that is, $N = A - Z$. Often in science books and journals the nuclear notation does not include Z and is written

$$^{A}X.$$

We can illustrate the notation using the most common form of carbon as an example. The chemical symbol is "C" and the nucleus contains six protons and six neutrons. The nuclear notation is

$$^{12}_{6}C \text{ or } ^{12}C.$$

Sometimes this form of carbon is written in the science literature as "carbon-12."

Because the number of protons (*Z*) identifies the element, there is a bit of redundancy in the notation where *Z* is specified, so the more common nuclear notation leaves the subscript off. However, for this chapter we will use the notation with both the superscript (*A*) and the subscript (*Z*) because they are both useful when trying to identify unknown nuclei in various nuclear processes.

There is another way to interpret *Z* that will be helpful in understanding the nuclear notation. As mentioned previously, the electric charge of the nucleus can be written in terms of *Ze*. Thus *Z* can be thought of as the charge in units of *e*. This idea will be important when we discuss other types of particles involved in nuclear decay, such as electrons ($Z = -1$), positrons ($Z = +1$), neutrons ($Z = 0$), and gamma rays ($Z = 0$), none of which are protons.

Isotopes and Stability

Isotopes (also called nuclides) are atoms that have different numbers of neutrons, even though they have the same number of protons. Isotopes of a particular atom all have the same atomic number *Z*, but different nucleon number or atomic mass number *A*. For example, carbon has several different isotopes, ranging from carbon-8 to carbon-22, which have six protons each, but different numbers of neutrons:carbon-8 has two neutrons, carbon-11 has five neutrons, and carbon-13 and carbon-14 have seven and eight neutrons, respectively. The nuclear notation for each of these isotopes of carbon is $^{8}_{6}C$, $^{11}_{6}C$, $^{13}_{6}C$, and $^{14}_{6}C$. Of the 15 isotopes of carbon that exist, only three, carbon-12, carbon-13, and carbon-14, occur naturally on earth; the others are produced artificially by nuclear reactions in the laboratory.

Carbon-14 is used for determining the age of old materials that are made up of organic matter. This technique, called radioactive dating, relies on the fact that a small amount of carbon-14 was present in the object while it was living, having taken in carbon dioxide during its lifetime. We will discuss the mathematics of radioactive decay in a later section.

Hydrogen provides another example of isotopes. It exists in three forms: hydrogen ($^{1}_{1}H$), deuterium ($^{2}_{1}H$), and tritium ($^{3}_{1}H$). All three have one proton, but hydrogen has no neutrons, deuterium has one neutron, and tritium has two neutrons.

Nuclei of isotopes can be stable or unstable. The stability of the nucleus of an atom is different from the stability of the electronic configuration of an atom or molecule. If all electronic energy levels are properly filled with electrons and the atom is in its ground state, the electronic configuration is stable. In a stable molecule, the electronic configuration favors the molecule remaining as it is. An unstable molecule will likely change its electronic configuration spontaneously, or it could split apart or react easily with other molecules or atoms with the slightest disturbance.

If the nucleus of an atom is stable, it will remain in its current state. If it is unstable, the isotope will emit radiation; it is said to be radioactive. Most elements have at least one stable isotope. For example, carbon-12 and carbon-13 are both stable isotopes of carbon, whereas all the other isotopes of carbon are unstable. Hydrogen and deuterium are both stable isotopes, but tritium is unstable. Elements with atomic numbers greater than 82 (lead) are all unstable. In addition, all of the elements beyond uranium, $Z > 92$, are artificially produced by nuclear reactions; they do not occur naturally. Isotopes that are unstable are often referred to as *radionuclides* or *radioisotopes*.

One of the reasons why some nuclei are unstable is that the nucleus is too large. For proton numbers less than about 20, there is about one neutron for every proton in the nucleus. For larger values of Z, the nucleus becomes larger so that the repulsive Coulomb force between protons becomes stronger than the strong nuclear force. Extra neutrons are needed to mediate the repulsive force by helping to bind the protons to other nucleons. Up to about $Z = 80$ to 82, the ratio of neutrons to protons increases to about 1.5. As Z goes above 82 to 83, the distance between protons on either side of the nucleus is large enough that the nucleus is unstable, and nucleons are ejected in the form of radiation.

There are some general rules to determine if nuclei with smaller mass numbers are stable or unstable. We won't go into all the details for these rules but will give some highlights. One rule is that for mass number less than 40 ($A < 40$), stable nuclei have about the same number of protons and neutrons. Another rule has to do with whether there is an even or odd number of protons and neutrons. Yet another one predicts stable nuclei and isotopes if the numbers of protons or neutrons are close to certain values, called magic numbers; this concept is similar to the filled electronic shells of atoms.

Types of Nuclear Radiation

When a nucleus is unstable, it emits nuclear radiation in the form of highly energetic particles and/or photons. The isotope is said to be radioactive and to undergo nuclear decay. There are several different types of nuclear radiation; we will describe three of these in more detail, specifically, *alpha*, *beta*, and *gamma decay*.

When a radioisotope emits radiation, the original radioactive nucleus is changed in some way. For example, in alpha decay the original radioisotope emits several nucleons, resulting in an isotope of a different element. In beta decay the numbers of protons and of neutrons change, again resulting in a new element. In gamma decay the energy of the original nucleus changes, but the numbers of protons and neutrons remain the same, resulting in the same isotope with different energy. The original nucleus is called the *parent* nucleus and the resulting nucleus is called the *daughter* nucleus.

To determine what the daughter nucleus is, we can write a *nuclear decay equation*. A nuclear decay equation looks like the following:

$$^{A}_{Z}X \rightarrow {}^{A'}_{Z'}X' + \text{radiation},$$

where $^{A}_{Z}X$ is the parent nucleus, $^{A'}_{Z'}X'$ is the daughter nucleus, and "radiation" is the particular type of radiation emitted by the parent nucleus; the radiation may be one or several particles or photons depending on the type of decay process. The arrow points in one direction from the parent to the daughter because the decay process goes in only one direction.

The nuclear decay equation is quite useful because there are two rules that must be followed. The first is that the number of nucleons A is conserved. That is, the total number of nucleons (protons and neutrons) in the parent must equal the number of nucleons in the daughter plus the number of nucleons in the radiation. The second is that the number of protons Z (or *charge* in units of e) is conserved. Thus, the number of protons or charge in the parent equals the number of protons or charge in the daughter plus the number of protons or charge in the radiation. Because the number of protons identifies the element, if Z and Z' are different (depending on the type of radiation emitted), the parent and daughter nuclei are different elements. However, if the radiation emitted by the parent does not result in a change in the number of protons, as in gamma decay, the parent and daughter nuclei will be the same element.

Alpha Decay

When a nucleus undergoes alpha decay, it emits an alpha particle. An alpha particle is identical to the nucleus of a helium atom consisting of two protons and two neutrons. The nuclear notation is ${}^{4}_{2}He$. Note that an alpha particle has a charge of $+2e$ because it does not contain the two electrons that are found in neutral helium (fig. 7.2).

Alpha decay occurs for plutonium-242. When this radioisotope

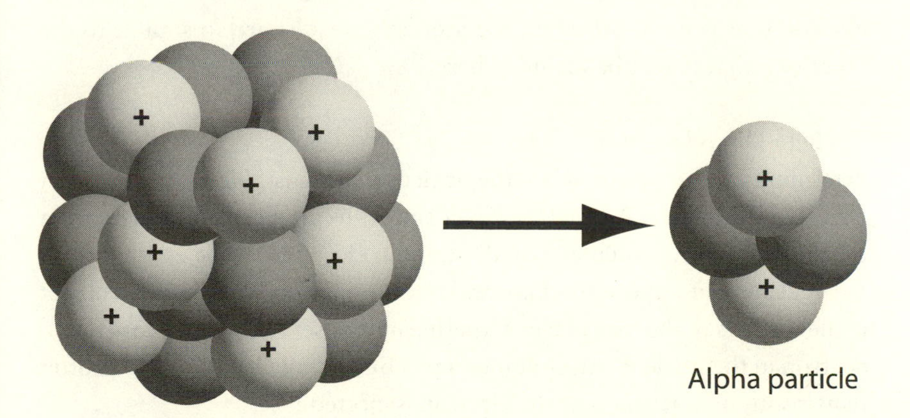

FIGURE 7.2. Alpha decay of a nucleus. It emits an alpha particle consisting of two protons and two neutrons, which is identical to the nucleus of a helium atom.

undergoes nuclear decay, it emits an alpha particle and the resulting isotope is uranium-238. The nuclear decay equation is

$$^{242}_{94}\text{Pu} \rightarrow {}^{238}_{92}\text{U} + {}^{4}_{2}\text{He},$$

where $^{242}_{94}\text{Pu}$ is the parent nucleus, $^{238}_{92}\text{U}$ is the daughter nucleus, and $^{4}_{2}\text{He}$ is the alpha particle. The decay equation obeys the two conservation rules: the number of nucleons is conserved (242 = 238 + 4) and the number of protons or charge is conserved (94 = 92 + 2).

Note that when a parent nucleus undergoes alpha decay A decreases by 4 and Z decreases by 2. Thus the parent and daughter are different elements. On the periodic chart, the parent and daughter are located near each other with only one element between them.

Radium-226 also decays by alpha decay to radon-222. The nuclear decay equation is

$$^{226}_{88}\text{Ra} \rightarrow {}^{222}_{86}\text{Rn} + {}^{4}_{2}\text{He}.$$

In the scientific literature, sometimes the Greek symbol for alpha (α) is used for the alpha particle.

Beta Decay

The discussion of beta decay will include three topics: (1) *beta-minus* (β^-) decay, (2) *beta-plus* (β^+) decay, and (3) *electron capture* (EC). EC is not actually a decay process: it is a situation where the nucleus captures an electron that is in an orbital of the atom; the results are so similar to the other two that it will be included here.

Beta-Minus Decay

Beta-minus decay occurs when the nucleus of an unstable isotope emits a beta-minus particle. A β^- particle is identical to an electron, which has a charge of $-e$. Now you may wonder how an electron can be emitted from the nucleus. But a more fundamental question is how did the electron get in the nucleus in the first place? Experimental observations indicate that a neutron in the nucleus can split into a proton and an electron. The proton remains in the nucleus, but the electron is ejected.

There is another particle emitted during beta-minus decay, and that is an *antineutrino*. *Neutrinos* are elementary particles that have very small mass and no charge, and move at very high speeds. Antineutrinos are examples of antimatter. Neutrinos (and antineutrinos) are important because they account for energy and momentum changes between the parent, the daughter, and the beta particle.

The nuclear decay equation for beta-minus decay is

$$ {}^{1}_{0}n \rightarrow {}^{1}_{1}p + {}^{0}_{-1}e + \bar{\nu}, $$

where ${}^{1}_{0}n$ is the nuclear notation for the neutron, ${}^{1}_{1}p$ for the proton, ${}^{0}_{-1}e$ for the beta-minus particle, or electron, and $\bar{\nu}$ for the antineutrino. The notation may be a bit surprising, so let's look at it more closely. Recall that the superscript A represents the number of nucleons (neutrons and protons), so the "1" for the neutron and the "1" for the proton are consistent. The subscript Z can be interpreted in two ways. One is that it represents the number of protons, so the "0" for the neutron makes sense, as does the "1" for the proton. For the electron, however, keep in mind that the subscript also stands for the charge of the nucleus as a multiple of e, the fundamental unit of charge. The electron has a charge of $-e$, so the "-1" makes sense using this definition, as do the subscripts for the neutron and the proton. Also, note that the two conservation rules are satisfied; the value for A is constant for the two sides of the equation $(1 = 1 + 0)$ and so is the value for Z $(0 = 1 - 1)$.

Carbon-14 is a radioactive isotope of carbon that decays by emitting a β^- particle. When the parent ${}^{14}_{6}C$ emits the β^- particle, the resulting daughter nucleus is nitrogen-14, or ${}^{14}_{7}N$. The nuclear decay equation is

$$ {}^{14}_{6}C \rightarrow {}^{14}_{7}N + {}^{0}_{-1}e + \bar{\nu}. $$

We see that the two conservation rules are obeyed: the number of nucleons remains constant at $A = 14$ on both sides $(14 = 14 + 0)$; and the value for Z is constant $(6 = 7 - 1)$. Note that the parent and daughter are different elements (carbon has six protons and nitrogen has seven). On the periodic table they are located next to each other, with the daughter having a higher number of protons.

Beta-Plus Decay

Beta-plus decay occurs when the nucleus of an unstable isotope emits a beta-plus particle. Another name for a β^+ particle is *positron*. A positron is very similar to an electron: it has the same mass, but has a charge of $+e$. A positron is an example of antimatter and is sometimes called an antielectron. Once again, you may wonder how a positron can be emitted from the nucleus. And again, experimental observations show that a proton in the nucleus splits into a neutron and a positron. The neutron remains in the nucleus, but the positron is ejected. A neutrino is emitted as well.

The nuclear decay equation for this process is

$$ {}_{1}^{1}p \rightarrow {}_{0}^{1}n + {}_{+1}^{0}e + v, $$

where ${}_{+1}^{0}e$ is the nuclear notation for the beta-plus particle, or positron, and v is the notation for the neutrino. For the positron, the subscript once again stands for the charge as a multiple of e, the fundamental unit of charge, which is +1. Again, note that the two conservation rules are satisfied; the values for A ($1 = 1 + 0$) and Z ($1 = 0 + 1$) are constant for the two sides of the equation.

An example of an isotope that decays by emitting a β^+ particle is oxygen-15. When the parent ${}_{8}^{15}O$ emits the β^+ particle, the resulting daughter nucleus is nitrogen-15, or ${}_{7}^{15}N$. The nuclear decay equation is

$$ {}_{8}^{15}O \rightarrow {}_{7}^{15}N + {}_{+1}^{0}e + v. $$

Again, notice that the parent and daughter are different elements with the atomic number decreasing by 1. The two conservation rules are obeyed: for A, $15 = 15 + 0$, and for Z, $8 = 7 + 1$. On the periodic table they are located next to each other, with the daughter having a lower number of protons.

Electron Capture

Electron capture is technically not a decay process, but it is so similar to beta-plus decay that it makes sense to describe it here. The basic idea of electron capture is that the nucleus captures one of the electrons that are in orbit around it. This electron combines with a proton to form a neutron. A neutrino is emitted by the nucleus during this process.

During electron capture the electronic configuration of the atom changes. Usually the nucleus captures an electron in a lower-lying energy state. Then an electron from a higher-energy state "falls" to the lower state. When it does, the electron has to give up this energy in the form of an electromagnetic (EM) wave. This EM wave is typically an x-ray; keep in mind that it is the atom that emits the EM wave and not the nucleus.

The nuclear equation for EC is

$$ {}^{1}_{1}p + {}^{0}_{-1}e \rightarrow {}^{1}_{0}n + v, $$

which shows that a neutrino is emitted. An example is beryllium-7 that captures an electron to become lithium-7. Its nuclear decay equation is

$$ {}^{7}_{4}\mathrm{Be} + {}^{0}_{-1}e \rightarrow {}^{7}_{3}\mathrm{Li} + v. $$

Notice that the parent and daughter nuclei differ by one proton in the same way as in β^+ decay, with the daughter, lithium-7, having one fewer proton than the parent. Once again, we note that the number of nucleons A $(7 + 0 = 7)$ and the charge number Z $(4 - 1 = 3)$ are conserved.

Recall that for β^+ decay a proton changes into a neutron (with the emission of a positron). For electron capture, a proton changes into a neutron (upon absorption of an electron) as well. Thus electron capture is said to be a competing process because there are two routes for the decay of the parent nucleus to the daughter nucleus. An example is that of cobalt-56. If it decays by emitting a beta-plus particle, then the decay equation is

$$ {}^{56}_{27}\mathrm{Co} \rightarrow {}^{56}_{26}\mathrm{Fe} + {}^{0}_{+1}e. $$

However, if electron capture occurs in cobalt-56, then the decay equation is

$$ {}^{56}_{27}\mathrm{Co} + {}^{0}_{-1}e \rightarrow {}^{56}_{26}\mathrm{Fe}. $$

In both cases, cobalt-56 is the parent nucleus, iron-56 is the daughter nucleus, and A and Z are conserved. (Note that a neutrino is emitted in both situations, but it is not shown in either decay equation above.)

Gamma Decay

Gamma decay occurs when the nucleus of an unstable isotope emits a gamma ray, a photon or electromagnetic wave that is highly energetic. Gamma ray emission occurs when the nucleus of an atom is in a high-energy state. When the nucleus changes to a lower-energy state, the ground state, a photon is emitted in the form of a gamma ray. The process is quite similar to that of an atom when an electron changes energy levels, jumping from a higher- to a lower-energy level. The energy of photons emitted from atoms because of electron transitions tends to be a few electron-volts in size, whereas the energy emitted by the nucleus is thousands or millions of electron-volts (keV or MeV).

When a nucleus emits a gamma ray, there is no change in the number of protons and neutrons; the nucleon number (or atomic mass number) A and the proton number (or atomic number) Z remain the same. Thus the parent and daughter nuclei are the same element.

To distinguish between the parent and the daughter in a nuclear decay equation, the parent is labeled in a slightly different way from the daughter. We can illustrate this with an example that is used in medical applications: technetium-99m. Technetium is useful in diagnostic procedures to determine if an organ of the body is functioning properly or if it may have an abnormality. An image of the organ is produced when gamma rays are emitted from technetium that is injected into the body and taken up by the organ.

The nuclear notation for radioactive technetium is ${}^{99m}_{43}\mathrm{Tc}$, where the "m" stands for "metastable." The nuclear decay equation is

$${}^{99m}_{43}\mathrm{Tc} \rightarrow {}^{99}_{43}\mathrm{Tc} + \gamma.$$

The parent is ${}^{99m}_{43}\mathrm{Tc}$ and the daughter is ${}^{99}_{43}\mathrm{Tc}$. Note that the gamma ray does not change the number of nucleons nor the number of protons. The nuclear notation can be written ${}^{0}_{0}\gamma$ but the subscript and superscript are not necessary. The energy of the gamma ray emitted from ${}^{99m}_{43}\mathrm{Tc}$ is 142.7 keV.

Decay Chain

Often, when an unstable nucleus emits radiation, the resulting daughter nucleus is unstable as well. The daughter nucleus becomes a parent nucleus and emits radiation in turn, producing a different daughter nucleus. This situation is called a *decay chain*. Sometimes the decay chain goes through many steps to reach a stable daughter isotope; other times there are only a few steps. The decay chain will be as long as it needs to be in order to reach a stable daughter isotope.

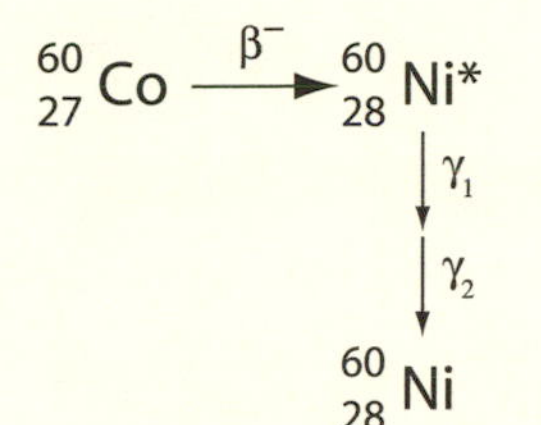

FIGURE 7.3. Decay of cobalt-60 to nickel-60 with emission of one beta-minus particle and two gamma rays. An antineutrino is also emitted but is not shown.

An example that is important in the medical field is cobalt-60, which is used to treat cancer by exposing cancerous tissue to the gamma rays that are emitted. The nucleus of cobalt-60 emits a beta-minus particle (and antineutrino) to become nickel-60; the nickel-60 nucleus is in an excited state before emitting two gamma rays to become a stable form of nickel-60 (fig. 7.3). The nuclear decay equation for the beta decay is

$$^{60}_{27}\text{Co} \rightarrow {}^{60}_{28}\text{Ni}^* + {}^{0}_{-1}e + \bar{\nu},$$

where the asterisk indicates an excited state of nickel-60. The decay equation for nickel-60 is

$$^{60}_{28}\text{Ni}^* \rightarrow {}^{60}_{28}\text{Ni} + \gamma_1 + \gamma_2$$

where γ_1 is the first photon whose energy is 1.17 MeV, and γ_2 is the second photon of energy 1.33 MeV. (The energy of the beta-minus particle in the first part is 0.31 MeV, or 310 keV.) Note that the parent is nickel and the daughter is nickel, and the asterisk is used to distinguish between them. The two decay equations can be combined so that the equation for the entire process is

$$^{60}_{27}\text{Co} \rightarrow {}^{60}_{28}\text{Ni} + {}^{0}_{-1}e + \gamma_1 + \gamma_2 + \bar{\nu},$$

which shows that the original radioactive cobalt-60 decays to nickel-60 by emitting a beta-minus particle and two gamma rays (and an antineutrino).

This is a two-step decay chain; another example illustrates a very long decay chain. Radioactive uranium-238 decays to stable lead-206 by going through 14 steps. Uranium-238 emits an alpha particle to become thorium-234, which emits a beta-minus particle to become palladium-234, and another beta-minus particle to become uranium-234. Uranium-234 emits an alpha particle to become thorium-230, which emits an alpha particle to become radium-226. The process continues until a total

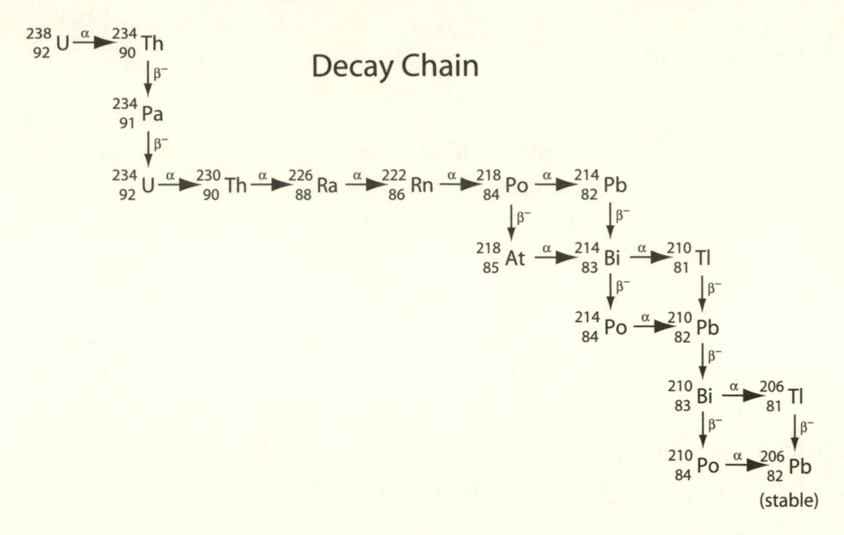

FIGURE 7.4. The decay chain of uranium-238 to lead-206. It involves the emission of eight alpha particles and six beta-minus particles. Antineutrinos are emitted during the beta-minus decay, but they are not shown.

of eight alpha particles and six beta-minus particles have been emitted, resulting in lead-206, which is a stable isotope. Recall that for each beta-minus decay an antineutrino is emitted as well. An interesting observation in this decay chain is that some of the resulting isotopes emit an alpha particle followed by a beta-minus particle, or a beta-minus particle followed by an alpha particle. The decay chain for this example is illustrated in figure 7.4.

The idea of a decay chain leading to many other radioactive isotopes brings up several major issues that have to be considered when using radioactive materials. For example, it may be that the only useful reaction for a particular application involves only one of the many steps in the decay chain. This could be because its energy has the right value, or because it lasts for the right length of time, or because the concentration is optimum so that correct results are obtained in the application. The rest of the decay processes in the decay chain may not be usable, so the problem of what to do with the "waste" material arises. Storage and handling of this waste material are important to minimize exposure to radiation that could continue for perhaps hundreds to millions of years.

Other Nuclear Processes

There are several other processes that involve the nucleus, such as other types of radiation, fission and fusion, and nuclear bombardment. Let's look at a few examples.

Neutrinos (mentioned previously) are uncharged particles that move very fast and are nearly undetectable. They are emitted in certain types of nuclear decay processes. They were first proposed in 1930 by Wolfgang Pauli because experiments involving beta decay appeared to violate energy and momentum conservation principles. It was found that neutrinos interact with nuclei though the weak nuclear force.

Neutron emission is a process where the nucleus emits a neutron. If the neutron strikes another nucleus, it can set up that nucleus in an excited state so it produces radiation (this is called *neutron activation*). An example of neutron emission is provided by californium-252, which is named for the state in which it was first made (at the University of California at Berkley). Californium-252 is an unstable isotope that decays by emitting a neutron. Its nuclear decay equation is

$$^{252}_{98}\mathrm{Cf} \rightarrow {}^{251}_{98}\mathrm{Cf} + {}^{1}_{0}n.$$

If the neutrons from this reaction strike a sample of nitrogen-14, the resulting nuclei are in an excited state, which will then emit a gamma ray. The decay equations are

$$^{1}_{0}n + {}^{14}_{7}\mathrm{N} \rightarrow {}^{15}_{7}\mathrm{N}^*$$

and

$$^{15}_{7}\mathrm{N}^* \rightarrow {}^{15}_{7}\mathrm{N} + \gamma.$$

Proton emission is a similar process where the nucleus emits a proton. Examples include cobalt-53 and thulium-147.

Nuclear fusion occurs when two nuclei join together to form another nucleus. It typically involves lighter nuclei such as hydrogen and helium. This process is an important mechanism in the production of light and heat from the sun. There is a series of fusion reactions that go on in the sun: the first is proton-proton (or hydrogen-hydrogen) fusion to produce deuterium; then a proton fuses with deuterium to produce helium-3; then two helium-3 nuclei fuse to produce helium-4. The overall effect is that six protons, or hydrogen nuclei, are involved in the fusion process to produce one helium-4 nucleus. During this nuclear cycle, there is a release of two positrons, two gamma rays, and two neutrinos, accompanied by the release of two protons and a large amount of energy.

Nuclear fission is a process where a nucleus splits apart into two different nuclei. An example is fission of plutonium-240, which splits into strontium-97 and barium-139 and emits four neutrons in the process. Its nuclear decay equation is

$$^{240}_{94}\text{Pu} \rightarrow {}^{97}_{38}\text{Sr} + {}^{139}_{56}\text{Ba} + 4({}^{1}_{0}n).$$

Mathematics of Radioactive Decay

A sample of radioactive material continually gives off radiation over a period of time, although the amount of radiation released per unit of time decreases as time elapses. Not all radioactive nuclei emit radiation at the same moment in time. Some kinds of radioactive material emit more radiation than others and some emit the radiation at a faster rate than others. The mathematics of radioactive decay helps us to quantify the amount of radiation emitted, so that we can use it properly in specific applications or determine the amount of exposure when handling it.

The radioactive decay of nuclei is a random process. If you were able to observe a single atom (or nucleus), you would not be able to say exactly when that nucleus would decay. You would, however, be able to determine the probability that the nucleus would decay over a certain period of time; still, it is just a prediction. For a collection of nuclei, there are statistical processes that are used to determine how many nuclei will decay, and a very well-defined mathematical formula is used to determine how many radioactive nuclei remain as a function of time. This formula is an exponential function.

To help us with the mathematics, let's look at two quantities, known as the *decay rate* and *half-life*. Then we will look closely at the mathematics of the exponential function as it relates to nuclear decay.

Decay Rate

When a sample of radioactive nuclei emits radiation, the number of decays that occur per unit of time is called the decay rate. Another name for decay rate is *activity*. Suppose that the number of radioactive nuclei is given by N, changing in time by ΔN. The decay rate, or activity, R is written

$$R = \left|\frac{\Delta N}{\Delta t}\right|, \tag{1}$$

where Δt is the time that elapses when ΔN nuclei change. (Ideally, Δt is taken as very small.) The absolute value in equation (1) indicates that even though the number of nuclei decreases, so that ΔN is a negative number, the decay rate is written as a positive value.

There are two common units used for decay rate. One is the SI unit of *becquerel* or Bq, which is the same as a decay/second. The other is the *curie*, or Ci, where 1 Ci = 3.70×10^{10} Bq. The becquerel is useful when talking about individual nuclei, whereas the curie is useful when discussing larger samples of nuclei. The becquerel is named for Henri Becquerel and the curie for Pierre and Marie Curie. Becquerel and the Curies were French physicists (Marie was of Polish descent and a chemist as well) who lived in the late 1800s and early 1900s. They were instrumental in the discovery of radiation and received the Nobel Prize in Physics in 1903 for their discovery. Later, in 1911, Marie Curie (also known as Madame Curie) received the Nobel Prize in Chemistry for the discovery of the elements radium and polonium. Madame Curie died in 1934 most likely due to exposure to radiation because the biological effects of radiation were unknown at the time. Pierre Curie had died many years earlier (1906) in an accident while crossing a street in Paris.

The decay rate depends on the number of parent nuclei that are present in the sample. Think about a very small sample of nuclei where only a few nuclei can change over a period of, say, a few minutes; if you had ten times as many nuclei, more nuclei could change over the same period of time. In fact, the decay rate is proportional to the number of parent nuclei and can be written

$$\frac{\Delta N}{\Delta t} = -\lambda N, \tag{2}$$

where λ is a constant of proportionality called the *decay constant*, and has different values for different radioactive isotopes (but is constant for any specific isotope). Note that the negative sign indicates that the number of parent nuclei is decreasing as time elapses.

Another way to illustrate that the decay rate depends on the number of nuclei present is to relate it to another rate quantity dealing with money. Suppose someone gave you $10 and insisted that you had to spend it over a two-day period. Your spending rate would be $10 divided by 2 days, or $5 per day. Now suppose you were given $100 to spend over two days; then your spending rate would be $50 per day. The more money you have, the greater your spending rate will be. The mathematical behavior of radioactive nuclei dictates that, for a certain type of nucleus, a certain percentage of nuclei must decay over a certain period of time. Thus, the more nuclei you start with, the greater the decay rate will be.

In calculus, the decay rate equation (2) can be written as a differential equation, which can be used to determine the number of radioactive parent nuclei (N, also called population) as a function of time. The result is

$$N = N_0 e^{-\lambda t}, \tag{3}$$

where N_0 is the initial number or population of radioactive nuclei in a sample at time $t = 0$. Note that this is an exponential function, with the minus sign indicating that it is a decaying and not a rising exponential. The formula shows us that the population of parent nuclei decreases with time from a starting population of N_0.

A plot of the exponential part of the population function [eq. (3)] is shown in figure 7.5. Note that the population decreases toward zero as time elapses. Theoretically the population never reaches zero (zero is an asymptote) no matter how large is the value for time. However, if the starting number of radioactive nuclei is small enough (say 100 or 1000 or 1,000,000), with enough time all the radioactive nuclei will decay. For larger samples of radioactive material, it may be easier to discuss the mass of the sample (more about this later).

The exponential part of equation (3) is sometimes called the *fraction remaining* because the quantity N/N_0 is the fraction of radioactive parent nuclei that remain as a function of time. This phrase will come up again when we discuss a medical application called MYOVIEW,[1] which uses

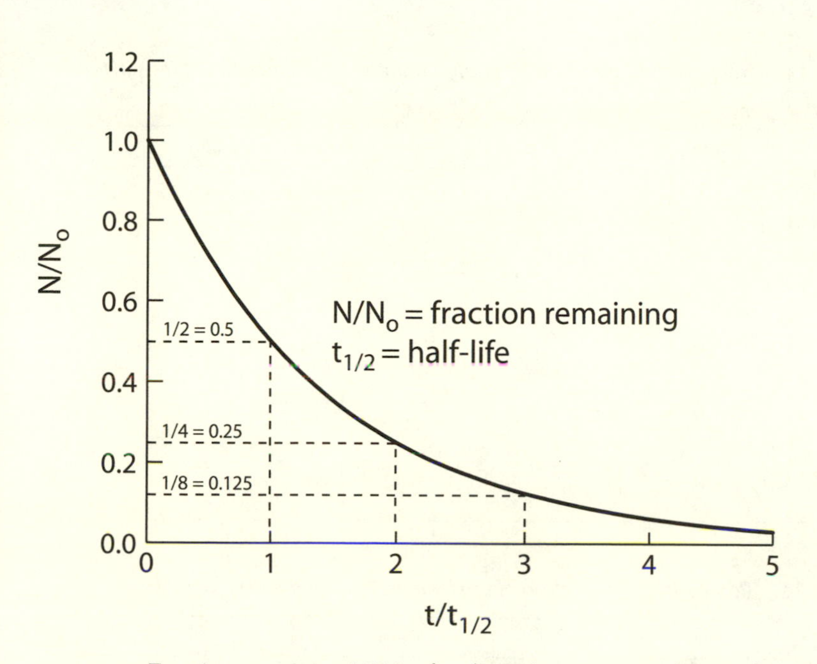

FIGURE 7.5. Fraction remaining N/N_0 of radioactive parent nuclei as a function of time. The fraction remaining corresponds to the exponential part of the decay function given in equation (3). The half-life $t_{1/2}$ is the time it takes for half of the nuclei to decay.

technetium-99m to form an image of the coronary arteries to help determine whether they are blocked.

Half-Life

The *half-life* is defined as the time that it takes for half of a sample of radioactive parent nuclei to decay. When a period of time equal to the half-life of a particular radioactive sample has elapsed, 50% of the sample will have decayed. If another half-life elapses, 50% of the remaining nuclei from the previous period will decay, so that only 25%, or 1/4, of the original number remains. The decay process continues so that only 12.5%, or 1/8, remains after three half-lives, 6.25%, or 1/16, after four half-lives, and so on (table 7.1).

Suppose we start with 1000 radioactive parent nuclei at time $t = 0$ and the half-life for the isotope is 5 seconds. Then after 5 seconds has elapsed, there will be 500 parent nuclei remaining; at 10 seconds, there

TABLE 7.1. *Fraction remaining of the initial population as a function of time t in terms of half-life $t_{1/2}$*

Time, t	$t/t_{1/2}$	Fraction remaining
0	0	$\left(\frac{1}{2}\right)^0 = 1$
$t_{1/2}$	1	$\left(\frac{1}{2}\right)^1 = \frac{1}{2} = 0.5$
$2t_{1/2}$	2	$\left(\frac{1}{2}\right)^2 = \frac{1}{4} = 0.25$
$3t_{1/2}$	3	$\left(\frac{1}{2}\right)^3 = \frac{1}{8} = 0.125$
$4t_{1/2}$	4	$\left(\frac{1}{2}\right)^4 = \frac{1}{16} = 0.0625$

TABLE 7.2. *Population as a function of time where $N_0 = 1000$ and $t_{1/2} = 5$ s*

Time (s)	Population N
0	1000
5	500
10	250
15	125
20	63

will be 250 remaining; at 15 seconds, there will be 125; and at 20 seconds, there will be 62 or 63 because the number of nuclei must be a whole number or integer (table 7.2).

The half-life of a radioisotope, written in symbol form as $t_{1/2}$, depends on the decay constant λ for that radioisotope. The mathematical relationship is

$$t_{1/2} = \frac{\ln 2}{\lambda} \tag{4}$$

or

$$\lambda = \frac{\ln 2}{t_{1/2}}, \tag{5}$$

where $\ln 2$ is the natural logarithm of the number 2. When λ is replaced in the exponentially decaying function, the function becomes

$$N = N_0 e^{-(\ln 2)t/t_{1/2}} \tag{6}$$

or

$$N = N_0 \left(\frac{1}{2}\right)^{t/t_{1/2}} . \tag{7}$$

A plot of the population as a function of time is shown in figure 7.6, with $N_0 = 1000$, $\lambda = 0.1386/\text{s}$, and $t_{1/2} = 5$ s.

Equation (6) once again shows the exponentially falling nature of the population as a function of time. Equation (7) is illustrated in table 7.1 for specific values of $t/t_{1/2}$.

Exponential Decay Functions for Mass and for Activity

Usually, it will not be known how many radioactive nuclei are present in a sample. However, the mass of the sample may be easily measured and it would be useful to have exponential decay functions written in terms of

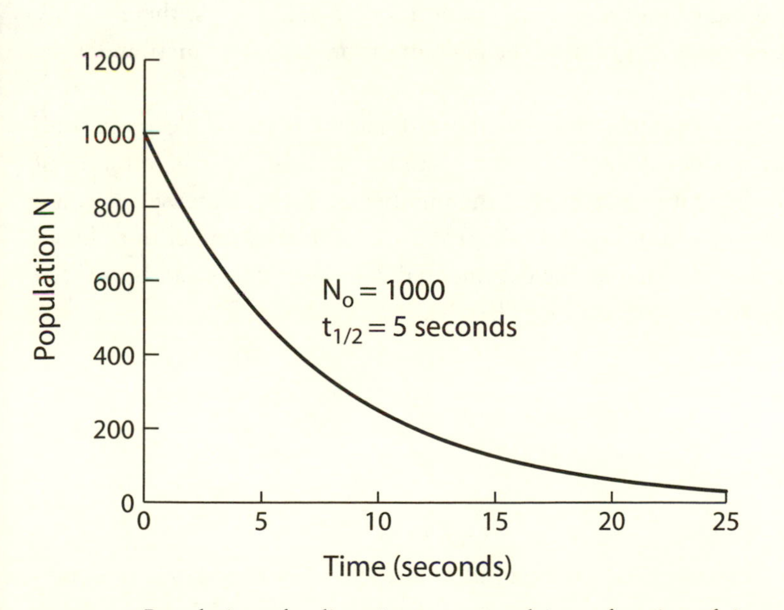

FIGURE 7.6. Population of radioactive parent nuclei as a function of time starting with a sample size, or initial population N_0, of 1000 nuclei and a half-life $t_{1/2} = 5$ s.

mass. This is quite easy to do when we realize that the mass of a sample is directly proportional to the number of nuclei present. Thus, equations (3) and (6) can be rewritten as

$$m = m_0 e^{-\lambda t}. \tag{8}$$

and

$$m = m_0 e^{-(\ln 2)t/t_{1/2}}, \tag{9}$$

where m_0 is the initial mass of the radioactive parent material at time $t = 0$ and m is the mass as a function of time.

Radon gas, which occurs naturally during the decay chain of uranium-238 (shown in figure 7.4), is an example. The isotopic form of radon formed during this process is radon-222, which is radioactive, decays by emitting an alpha particle to become polonium-218, and has a half-life of 3.825 days. Radon gas is a particularly important radioisotope because it is often found in houses in the United States. It can seep into a house through cracks and holes in the foundation or in basement walls. Suppose we have a 1.00 mg sample of radioactive radon-222. After 3.825 days, only 0.50 mg of radon remains; after another 3.825 days, there is only 0.25 mg of radon. A graph of the mass of radon-222 as a function of time is shown in figure 7.7.

Not only does the mass of the radioactive material decrease with time, but so does the activity of the sample. In other words, the rate of decay (the activity) decreases as the number of nuclei decreases. Because the activity of a sample is directly proportional to the number of radioactive nuclei in the sample, the exponential decay equations can be written for the activity of the sample. They become

$$R = R_0 e^{-\lambda t} \tag{10}$$

and

$$R = R_0 e^{-(\ln 2)t/t_{1/2}}, \tag{11}$$

where R_0 is the initial activity at time $t = 0$.

Let's consider an example used in medicine. Sodium iodide is a compound that normally uses the stable isotope iodine-127. However, if iodine-127 is replaced with radioactive iodine-131, the resulting com-

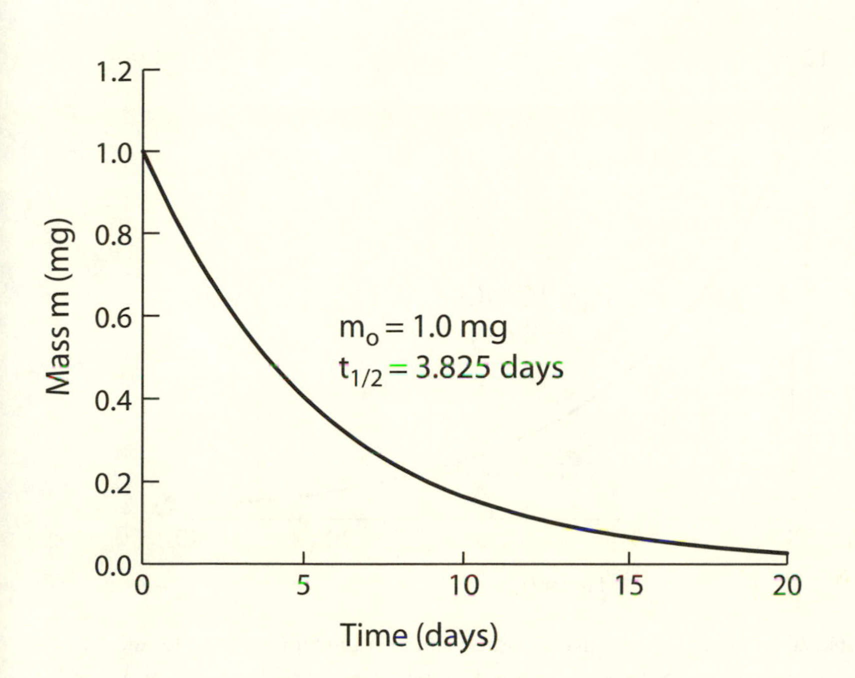

FIGURE 7.7. Plot of the mass of radioactive radon-222 as a function of time starting with an initial sample of mass m_0 = 1.0 mg and a half-life $t_{1/2}$ = 3.826 days.

pound is radioactive. The thyroid gland contains relatively high concentrations of iodine, so if the radioactive sodium iodide enters the patient, it can be used to treat hyperthyroidism (an overactive thyroid gland) and thyroid cancer. Iodine-131 decays by emitting a beta-minus particle and has a half-life of 8 days. The medication is taken in the form of a capsule that emits about 4 to 10 mCi (millicurie) of radiation. To illustrate the mathematics of the decay rate, suppose we have a capsule that has an activity of 10 mCi. After 8 days, the activity will be 5 mCi, after 16 days, the activity will be 2.5 mCi, and so on. The physician who administers the dose may wonder how long it will take for the activity to decrease to only 1 mCi, or 10% of the starting activity. If these numbers are used in the exponential decay equation (R_0 = 10 mCi, R = 1 mCi), it can be shown that it takes almost 27 days for the activity to be at that level. A plot of the activity of iodine-131 used in this example is shown in figure 7.8.

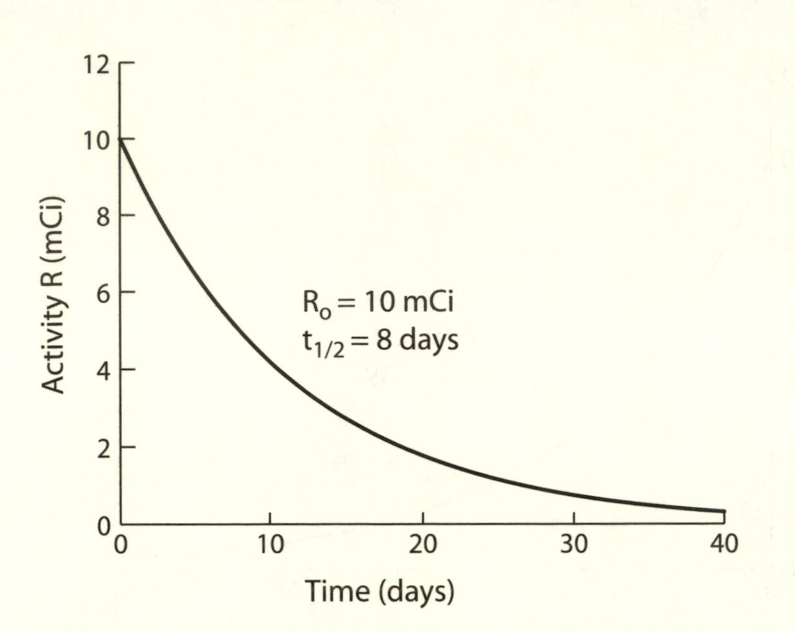

FIGURE 7.8. Activity of radioactive iodine-131 as a function of time. Iodine-131 emits a beta-minus particle and is used to treat hypothyroidism. The initial activity R_0 of the sample is 10 mCi, which is a typical dose given to a patient. The half-life of iodine-131 is $t_{1/2}$ = 8 days.

Effects of Radiation in Humans

Damage to human tissue can occur when it is exposed to nuclear radiation. The energy of alpha particles, beta particles, gamma rays, and x-rays is high enough to cause damage if it is absorbed. Usually the damage occurs when the radiation causes ionization of a large number of atoms that form the tissue. This is where the term "ionizing radiation" comes from. When a particle or ray strikes an atom, it can remove one or more electrons producing an ion. If enough of these ions are formed, then damage can occur to large molecules like DNA, to cells, and on an even larger scale to the tissue that has been exposed. If the damage occurs to healthy tissue, then significant problems can arise, such as development of cancer. On the other hand, if cancerous tissue is exposed to ionizing radiation, the cancer may be destroyed.

Damage occurs when the energy of the particle or photon is transferred to the medium in which it moves. Recall that the energies are quite large, on the order of hundreds of keV or MeV. Oftentimes, enough energy is available from just one particle of radiation that many ions can be formed. However, sometimes the radiation is not absorbed and will pass through the tissue without causing harm. The variation in absorption of the particles or photons depends on the type of radiation emitted.

Recall that an alpha particle is identical to the nucleus of a helium atom, with two protons and two neutrons. Alpha particles have a large mass compared to the other types of radiation. If we think of the formula for kinetic energy

$$K = \tfrac{1}{2}mv^2, \qquad (12)$$

we see that alpha particles do not move at high speeds compared to beta particles and gamma rays. For example, if an alpha particle has kinetic energy of 10 MeV, it moves at about 2.2×10^7 m/s (less than one-tenth the speed of light $c = 3 \times 10^8$ m/s). Alpha particles have a charge of $+2e$ because they lack the two electrons needed to be a neutral atom. With this relatively large charge, alpha particles interact strongly with materials around them. In fact, if an alpha particle is moving in air, it travels only a few centimeters before it is stopped. Even a sheet of paper could stop alpha particles.

Beta particles, either beta-minus (electron) or beta-plus (positron), have a smaller amount of charge, either $-e$ or $+e$. They also have a much smaller mass, over 7000 times smaller than the mass of an alpha particle, so for the same kinetic energy, they travel much faster than alpha particles. In fact, equation (12) is not applicable for beta particles because they move at relativistic speeds (very close to the speed of light c). Because of these two conditions (smaller charge and higher speed) it takes a bit more to stop beta particles than alpha particles. The average beta particle can travel several meters through air before it is stopped, or it can be stopped by a few millimeters of aluminum.

Gamma rays, as well as x-rays, are uncharged particles or photons, with no mass. They move at the speed of light because they are electromagnetic waves. They are not easily stopped, but can be stopped by a centimeter or more of lead.

Different types of radiation can penetrate human tissue and cause different amounts of damage. Alpha particles do the most damage; beta particles do a small amount of damage; and gamma rays and x-rays do very little damage (but still some).

You may think that alpha particles would not do much damage because they can be easily stopped. However, they actually do the most damage because they are so easily absorbed. They penetrate only a fraction of a millimeter into human tissue, so the energy of an alpha particle is dumped into a relatively small region of space. Many electronic bonds are broken, producing many ions, resulting in a lot of damage to the tissue. If alpha particles strike the skin on the surface of the body, the result can be severe radiation damage. At nuclear reactor facilities, there is quite a lot of shielding required because of the high production of alpha particles from radioactive isotopes with large nuclei, such as plutonium and uranium.

You may recall the case of a Russian spy who died in 2006. The individual was "poisoned" by ingesting radioactive polonium-210, which decays by emitting an alpha particle. Because alpha particles are so easily absorbed, exposure from outside his body would have caused serious damage to the skin, but the damage would have been limited to outer tissues. However, because the material was sprinkled on his food, his stomach and other organs were exposed, causing so much internal damage that it killed him.

Beta particles are fairly easily absorbed and penetrate a few millimeters into human tissue. Some damage occurs, but not nearly as much as with alpha particles. Some shielding is required to minimize exposure to radiation at reactor sites or nuclear medicine facilities.

Gamma rays and x-rays are not easily absorbed by human tissue, and most pass straight through the body. In the denser or thicker regions of the body, there is a greater chance of them being absorbed. This is evident if you look at an x-ray of a broken bone, or one taken at the dentist's office. The bone is much denser so that more x-rays are absorbed than in the surrounding tissue. However, depending on the density and thickness of the surrounding tissue you can see where the x-rays have been absorbed in those regions as well. Actually an x-ray is useful because those rays that are not absorbed by the body will pass through it and strike a photographic plate, exposing it much like film in a camera. The resulting image is a "negative" where the lighter areas, such as where a bone is located, have the least exposure, and the darker areas have the most exposure.

Because gamma rays and x-rays are ionizing radiation, if they are absorbed by the body, they can produce ions that have the potential to cause damage to tissue. Thus, it is important to minimize exposure to x-rays and gamma rays in order to minimize damage. X-ray technicians usually step behind a lead wall or other barrier to limit their exposure. Also, a lead apron is placed over a patient's chest and abdomen in order to block x-rays from passing through vital organs when it is not necessary.

Radiation Dosage

When a person is exposed to nuclear radiation, he or she will absorb some of it. It is important to know how much is absorbed so that the amount can be properly controlled, either to limit exposure to good tissue or to provide the proper exposure to cancerous tissue. The amount of radiation absorbed by human tissue is quantified in two slightly different ways. The first method is called *dose* and the second is called *effective dose*. The first method is related to how much energy per mass is deposited when the radiation is absorbed. The common unit is the *rad*, which stands for *radiation absorbed dose*, and is equal to 0.01 J/kg. The SI unit for dose is the *gray* (Gy), which is equal to 1 J/kg or 100 rad.

However, recall that alpha particles do the most damage to tissue and beta particles don't do nearly as much. Gamma rays and x-rays do about the same amount of damage, but it is very little compared to the others. Experiments have shown that alpha particles do about 20 times more damage to human tissue than gamma rays and x-rays. Beta particles do slightly more damage than gamma rays and x-rays, about 1.2 to 1.5 times more. To account for these differences, the second method for measuring dosage is used, the effective dose. This is found by taking the dose of a particular type of radiation and multiplying it by the damage factor, which is called the *relative biological effectiveness*, or RBE. Mathematically, it is expressed as

$$\text{Effective dose} = \text{dose} \times \text{RBE}.$$

TABLE 7.3. *Average effective radiation dose of US residents*

Source of radiation	Dose (millirem/year)
Natural background radiation	
Cosmic rays	28
Radioactive earth and air	28
Internal radioactive nuclei (^{14}C, ^{40}K)	39
Inhaled radon	~200
Man-made radiation	
Consumer products	10
Medical and dental diagnostics	39
Nuclear medicine	14
Total	~360

The effective dose is usually measured in units of *rem*, which stands for *radiation effective man*. The SI unit for the effective dose is the *sievert* (Sv), where 1 Sv = 100 rem.

The effective dose is utilized because it quantifies the amount of damage inflicted independent of the type of radiation. In other words, *1 rem of any type of radiation does about the same amount of damage to human tissue*: 1 rem of alpha radiation does the same damage as 1 rem of beta radiation and as 1 rem of gamma or x-ray radiation.

Using effective dose, it is easier to establish exposure criteria for safety reasons. Maximum exposure values are recommended to be 5 rem over a period of a year and 3 rem over a three-month period. A typical exposure for U.S. citizens is about 400 millirem per year. This exposure comes from a variety of sources, including cosmic rays, naturally occurring radioisotopes, and man-made sources such as consumer products and medical applications. The largest source of radiation exposure likely comes from inhaled radon gas, which varies widely by geographical region. Table 7.3 gives more specific values of the sources of radiation.[2,3]

Radiation sickness occurs when a person is exposed to excessive amounts of radiation. Symptoms include nausea, fatigue, vomiting, hair loss, diarrhea, and bleeding when one is exposed to an effective dosage of 50 to 100 rem at one time. Exposure to about 400 rem will result in death within about 4 months. Exposure to 1000 rem will cause death within a few days or weeks. Exposure to 2000 rem will result in unconsciousness within minutes followed by death within a few hours.[4,5] (Note: For beta radiation, the dose in rad is about the same as the effective dose in rem, while for gamma rays and x-rays, the dose in rad is exactly the same as in rem. In medical applications, most exposure is to gamma rays, x-rays, and beta particles, rather than alpha particles, so sometimes the exposure rates are given in rad instead of rem.)

Exposures in various medical procedures vary widely. Typical doses for x-ray procedures (radiology exams) range from 1 to 6 mrem for an x-ray of the skull or chest or limbs and joints, about 0.4 mrem for a dental x-ray, and 30 to 80 mrem for x-rays of the spine or abdomen or hips. CT (*computed tomography*) scans, which are x-rays, have effective doses of 200 to 1100 mrem, and a typical mammogram about 13 mrem. Various imaging techniques involving gamma rays or beta particles range from 150 mrem for a lung scan to 1700 mrem for a heart scan depending on the type of radionuclide used. The Health Physics Society has a great deal of information on exposure levels at its website.[6]

Medical Applications

Because of all the problems that nuclear radiation causes if absorbed by the human body, it is wise to avoid exposure if possible. However, nuclear radiation is used for a variety of medical techniques, such as treating illness or disease (therapeutic applications) by destroying cancerous tissue, or determining if an organ or part of the body is functioning properly (diagnostic applications).

Therapeutic Techniques

In the treatment of cancer, there are many applications of nuclear radiation, specifically gamma rays and beta particles. During these treatments, radiation strikes the cancerous tissue in order to stop the rapid production and growth of cells by changing their chemical and/or physical properties.

Exposure of cancerous tissue to radiation is accomplished by using either external or internal devices. External devices typically use gamma emitters so that the gamma rays can pass through to the target area without much absorption by healthy tissue. Cobalt-60 is primarily a source of gamma rays (even though beta-minus particles are produced). Special

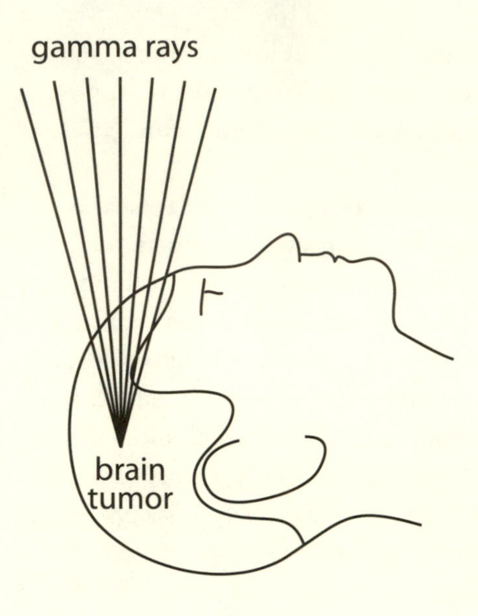

FIGURE 7.9. The Gamma Knife uses many low-level beams of gamma rays that intersect at the location of a tumor to deliver a high-level dose of radiation.

devices like the *Gamma Knife* and the *CyberKnife* use multiple beams of gamma rays that intersect at precise points to produce a region of high intensity (fig. 7.9). Each individual beam has a relatively low intensity so that not much damage occurs in healthy tissue, but as many as 100 other beams intersect to a region of just a few millimeters to give a high intensity. These devices are especially useful for treating brain tumors so that surgery (a very invasive technique) may not be required.

Internal devices can be used to treat cancer as well. They are usually beta emitters because beta particles are more easily absorbed than gamma rays. A procedure known as *brachytherapy* uses radioactive material that is embedded in small objects called seeds or beads. These objects are surgically implanted in or near a tumor so that they can be absorbed by the cancerous tissue.[7]

Another procedure is called radioimmunotherapy, or RIT, where a drug is tagged with a radioisotope and the drug targets a specific organ or system in the body. They are usually beta emitters. Examples include yttrium-90, which is used to treat non-Hodgkin's lymphoma, and iodine-131, which is used to treat thyroid cancer.[8]

Certain drugs can be used to treat pain; these are called palliative agents. Strontium-98 and samarium-153 are both beta emitters that are used to relieve pain due to bone disease.

There are many other procedures used to treat a variety of health conditions. Some are well established and others are still experimental. One such procedure is used in treatment of heart disease. When a coronary artery becomes blocked, the flow of blood that supplies proper nutrients to the heart muscle is decreased. If there is not sufficient flow, a heart attack can occur. Several treatments for this condition have been developed, including bypass surgery. A more recent technique is balloon angioplasty, in which a catheter is inserted into an artery in the leg and manipulated so that it goes to the coronary arteries. Once the catheter is in place, a small, thin balloon is inflated to help open up the artery. Also, a device called a stent is attached to the catheter and is usually left in the artery to keep the artery open when the catheter and balloon are removed. The stent is a flexible, metal wire cage (similar to chicken wire) that maintains its shape after being opened up and put into position with the catheter and balloon. After the surgery, over time the artery will often reclose (this is called *stenosis*) at the ends of the stent. If a radioisotope is embedded in a catheter near the ends of the stent, the radiation emitted exposes the plaque in the arteries to help prevent the arteries from reclosing by inhibiting the growth of cells and their migration.[9]

Diagnostic Techniques

A number of diagnostic techniques use devices that detect radiation in order to determine if a part of the body is functioning properly. Diagnostic applications primarily involve radioactive drugs or radiopharmaceuticals that target a particular organ or body system. These devices detect radiation emitted by the drugs, and then form an image of the target organ or system. This imaging process is called tomography.

As in other medical applications, the radiation used in these techniques is either gamma rays or beta particles. One technique is called single-photon emission computed tomography or SPECT, which measures gamma radiation. Another is called positron emission tomography (PET), which measures positron (beta-plus) radiation. The PET scan is used primarily for detection of tumors and diagnosis of brain disease. When x-rays are emitted after electron capture, certain types of x-ray detectors can measure them, and electron detectors, or beta-minus detectors, are used in imaging as well.

Other more familiar techniques include the standard x-ray machine used in dentists' offices, to look for broken bones, or in mammography. Computed tomography (CT) or "cat" scans use x-rays as well.

A technique that I am quite familiar with, having had this procedure done recently, uses a SPECT machine for a nuclear stress test. The test is based on the use of a special drug to form an image of blood flow through the coronary arteries. The procedure uses a MYOVIEW kit containing the drug Tetrofosmin, which has technetium-99m as the active ingredient. You may recall that technetium-99m emits gamma rays. After I walked on a treadmill for a while to get my heart rate up, the Tetrofosmin was injected into the bloodstream, and the drug was distributed throughout my body in just a few short minutes. Then I lay down on the SPECT machine which consisted of two large detector panels that measured the gamma rays being emitted from my chest. The computer then produced an image of the coronary arteries to see if proper flow of blood was occurring—I'm thankful to say that it was!

An information sheet that comes with the kit includes a description of the physical characteristics of radioactive technetium. The energy of the gamma rays is listed as 140.5 keV and the half-life as 6.03 hours. The sheet also describes the effectiveness of lead shielding, and I recall that the syringe containing the Tetrofosmin was sitting in a lead cup while I was on the treadmill. In addition, the information sheet shows a decay chart that lists the fraction of remaining material as a function of time based on the half-life of ^{99m}Tc.[10] You should recall that the "fraction remaining" is the just exponential part of the exponential decay equation.

Table 7.4 lists the information from the decay chart in the MYOVIEW information sheet. To illustrate the usefulness of the chart, let's look at the following example. Suppose a patient is injected with 0.320 mg of Tetrofosmin. The decay chart shows that after 2 hours time has elapsed the fraction remaining is 0.795 of the starting amount. Thus there should be about 0.795×0.320 mg = 0.254 mg of radioactive Tetrofosmin remaining in the patient. There may be other ways that the drug could leave the body, such as through urination, but the fraction-remaining calculation assumes that there is no other path of elimination from the patient in that time. For times greater than 12 hours, the fraction remaining can be calculated from the other values given in the chart. For example, to determine the fraction remaining at 14 hours, the fraction at 12 hours is

TABLE 7.4. *Decay chart for technetium-99m from a MYOVIEW kit*

Hours*	Fraction remaining	Hours	Fraction remaining
0	1.000	7	0.447
1	0.891	8	0.399
2	0.795	9	0.355
3	0.708	10	0.317
4	0.631	11	0.282
5	0.563	12	0.252
6	0.502	24	0.063

*Calibration time measured from the time of preparation.

multiplied by the fraction at 2 hours: $0.252 \times 0.795 = 0.200$. Thus, the patient should have 0.200×0.320 mg = 0.064 mg of drug after that time, again assuming no excretion.

Other Uses of Radiation

There are many other uses of nuclear radiation. It is used in nuclear power plants to generate electrical energy: when certain nuclei emit radiation, enough heat is produced to boil water, which gives off high-pressure steam, and this can be used to turn electrical turbines. It is used in carbon-14 dating, which is important to determine the age of dead organic matter—the half-life of carbon-14 is 5730 years, so it can be used to determine ages up to about 60,000 years. Another use is food irradiation in order to kill microorganisms and bacteria that can be harmful for human consumption—usually gamma rays are used, but x-rays and beta particles can be used as well.

Summary

We live and move and breathe in a world where we are exposed to nuclear radiation. The effects of nuclear radiation on the human body vary according to the amount of exposure. With very little exposure there are almost no noticeable effects, but with significant exposure severe damage can be done. While exposure to radiation can cause health problems and even cancer, radiation can be used to treat cancer as well. There are many people in the healthcare industry working to provide appropriate diagnostic and therapeutic procedures using radiation, from the technician who takes x-rays to the pharmacist who administers a radioactive tracer and the medical physicist who measures dosage amounts based on beam intensity and size. A good understanding of nuclear radiation and its biological effects is needed in this important area of medicine.

EIGHT

Drug Delivery and Concentration

What is a chapter on drug concentrations doing in a book on physics of the human body? While this is not really a physics topic—in fact some would say that it is more closely related to chemistry—it is a practical example of the types of problems that physicists often like to approach. When physicists look at a physical situation, they often try to apply mathematical models in order to quantify certain characteristics of the situation and to make predictions about the behavior of systems when certain parameters are known. An understanding of drug concentrations in the body, particularly how they change over time, can help scientists and health professionals when studying and/or treating diseases, injuries, or other health problems.

How are injections (shots) different from oral medication? How much medication is needed to treat various medical conditions? Why does someone need to take multiple doses of a drug rather than just one? In this chapter we will look at ways that drugs enter the body, how quickly they are absorbed by the body, and reasons why their concentrations change with time as they are used or removed (excreted). There will be a good bit of mathematics, but I will explain the physical basis for the mathematical models used. I will conclude this chapter with a brief look at some applications such as time-release medication, measurement of blood alcohol content, and drug testing.

Definitions

Several terms that I use in this chapter should be defined. They may not be new, but have different meanings when used in the context of drugs.

The study of the time-dependent behavior of drug concentrations in the human body is called *pharmacokinetics*. (The word *kinetics* refers to motion or to rate of change.) There are three processes that play a role in the changing drug concentrations: absorption, distribution, and elimination. *Absorption* occurs when the drug enters the bloodstream; it can occur rapidly such as when an injection or shot is given to a person directly into a vein, or slowly such as when a drug patch is worn on the surface of the skin, or at an intermediate rate such as when oral medication is taken. Absorption can take place in various tissues, much as a sponge absorbs water; but in the context of this chapter, *absorption refers to the drug entering the bloodstream*. Sometimes the phrase *systemic absorption* is used. Just a brief side note here: when a drug is given that goes immediately into the vein, like an injection, pharmacokineticists do not use the term absorption, but rather that the drug is available for immediate use. For our discussion in this chapter, I will use the more general definition as stated above.

Distribution of drugs occurs in the circulatory system as the movement of blood carries the drug to various types of tissues in the body. *Elimination* of drugs occurs through (1) excretion, where the drug is passed along with waste products in urine, feces, or sweat, and (2) metabolism, where the drug is broken down to an inactive chemical product either in the target tissue, or in the liver, the stomach, or the kidneys.

The phrase *administration of a drug* applies to the act of giving a drug to a person. A variation uses the word *administer*, which is the action of giving medication to a person, as in "A drug is administered to a person." The *route of administration* means the method used to give the drug, for example, by mouth (orally) or by injection (intravenously).

Pharmacological effect is a term that means the outcome or clinical effect of a drug. Some effects are intended or desired: decreased pain when taking medicine for a headache, shrinkage of a tumor when undergoing chemotherapy, or alleviation of congestion when taking cold medicine. Some effects are not intended nor desired (side effects): stomach discomfort when taking pain medication, loss of hair from drugs taken for cancer, and difficulty in urinating when taking an antihistamine or a decongestant.

The term *bioavailability* refers to the systemic availability (i.e., in the bloodstream) of a drug to be used by the body to produce a desired pharmacological effect, not to the availability of the drug on the market as a consumer product to be purchased. The bioavailability may depend on the dose of the drug, the frequency with which it is given, the route of administration, the form of the drug, and the actual drug product itself (as made by different manufacturers).

There are many other terms that will come up in this discussion, such as *bolus*, *toxicity*, and *first-order process*, but I will define them as they are introduced.

Methods of Drug Administration

Let's look first at the ways that drugs can enter the body. All of these methods increase the concentration of drugs in the body, but some are safer or more popular, whereas others are needed for faster results.

Taking medication orally is probably the first thing that comes to mind if you have a headache or a bad cough. Tablets and capsules are easy to put into a bottle, stock in a store, and purchase. And they are relatively easy to swallow when needed. Cough syrups and other liquid medications are convenient as well. The oral method is quite safe and does not require the presence of a medical professional to administer. The bioavailability of a drug taken orally varies depending on how easily it can pass through the stomach and/or intestinal lining (the gastrointestinal lining or *GI wall*) and be absorbed by the bloodstream (fig. 8.1). The rate of absorption tends to be quite fast for liquid medicine as well as for tablets and capsules once they dissolve. Many drugs cannot be taken orally

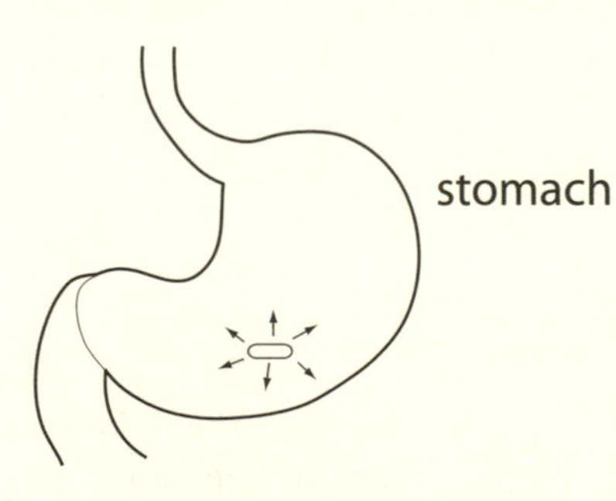

FIGURE 8.1. Oral medication dissolves in the gastrointestinal (GI) tract. It must pass through the lining of the stomach and intestines to be absorbed by the bloodstream.

because they may be unstable or may be metabolized by the liver before being absorbed in the bloodstream.

Another method that makes some people a bit squeamish is the dreaded shot, or injection. In this method, medication in the form of a liquid is loaded into a syringe with a needle attached to it. The needle is inserted into the arm, hip, or other exposed area, and the liquid is injected into the area.

There are several different ways that injections are given. The most effective method is by injection of the medicine directly into a vein of the circulatory system. This type of injection is called an *intravenous bolus*, or *IV bolus* (fig. 8.2). In this method the drug is available for immediate use because absorption is instantaneous (or nearly so) and 100% of the drug is in the bloodstream. However, problems may arise suddenly such as a bad reaction or a severe side effect.

The *intramuscular injection* is probably the most common injection method, as when getting a flu shot or other vaccines. The medication is injected into muscular tissue, which results in fairly rapid absorption into the bloodstream, particularly if the solution is water based. Intramuscular injections are easier to administer than an IV injection. Problems tend to be related to pain in the muscle tissue, particularly if the drug irritates the tissue.

The *subcutaneous injection* occurs by injecting medication into the fatty tissue layer beneath the surface of the skin. It is the preferred method where injections have to be given often, perhaps daily or more often, such as insulin shots for someone who has diabetes. Usually only small amounts of medication are needed and the needle is fairly short, causing

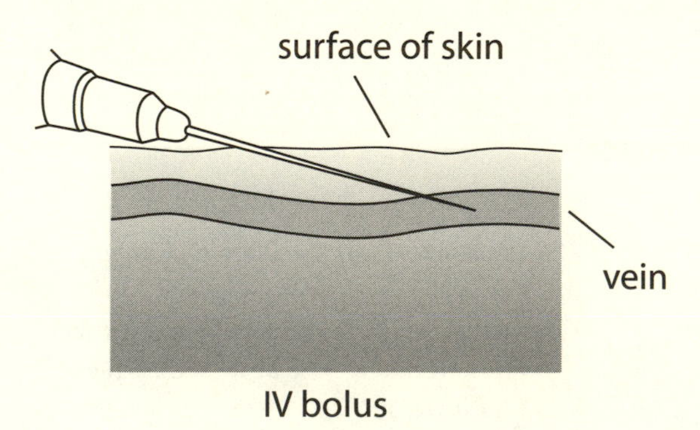

FIGURE 8.2. The IV bolus is an injection into a vein.

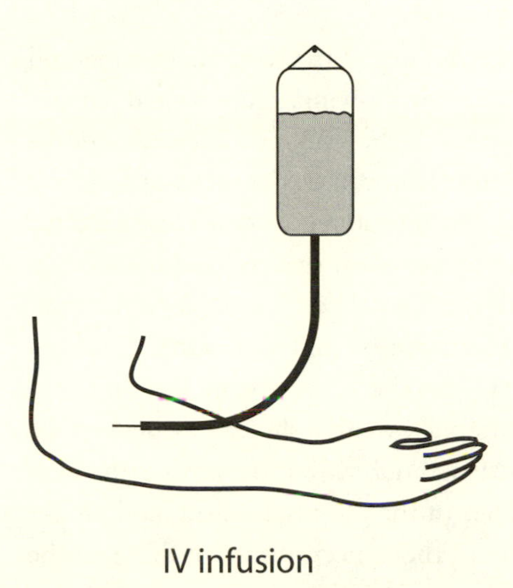

FIGURE 8.3. The IV infusion flows into a vein at a certain rate, determined by the height of the bag and control equipment placed in the tube leading from the bag to the needle.

very little, if any, discomfort. Subcutaneous injections are quite easy to self-administer. Absorption is not as quick as for intramuscular injections and may take several hours.

Another method of drug administration is *intravenous infusion*, or IV infusion (fig. 8.3). You may have seen this method in a hospital—medication in an IV bag hanging from a pole is attached to a tube, which is connected to a relatively large needle that is inserted into a vein in the arm or hand. The rate of infusion is determined by the flow rate in the tube, usually established by a flow meter as long as the bag is suspended from a high enough position (see chapter 2 for an example of flow rate). This method is very effective because the medication is immediately available; it is rapidly and completely absorbed (or available) just as in the case of the IV bolus (injection). The rate of absorption (or availability), however, is different because it depends on the infusion rate. This method is particularly useful for controlling drug concentration more precisely.

There are several other methods of drug administration, most of which occur when the drug passes through skin, mucous membranes, or other tissue. Buccal or sublingual drugs are placed in the mouth either between the cheek and gum (buccal) or under the tongue (sublingual). Liquids and fast-dissolving tablets such as lozenges work best and are absorbed into the bloodstream through the mucous membranes. Rectal drugs are placed in the rectum in the form of a suppository or a cream. Absorption of the drug occurs through the mucous membranes of the rectum. This method is useful for a patient who has difficulty swallowing. Transdermal patches or topical ointments are applied directly to the skin. Transdermal absorption tends to be slow so that the medication can be delivered over a long period of time up to several days. Topical ointments are usually used for local treatment of skin problems such as a rash or acne, but absorption into the bloodstream can occur. Nasal drug delivery can be used for local treatment of the nasal membranes (nasal sprays and nasal drops), but can also be used to provide drug delivery into the bloodstream. A flu mist injected through the nose is available in place of a flu shot. Drug delivery by inhalation is common for people with asthma, with rapid absorption over a large surface area of the membranes of the mouth, trachea, and lungs.

Later we will look more closely at the concentration of drugs in the body as it changes with time, and how different methods give different results. In particular we will look at IV injection and IV infusion, both of which involve immediate absorption of the medication. In addition, we will look at what happens when medication is taken orally. As it turns out, all of the other methods mentioned are similar in behavior to oral administration because the drug must be absorbed by diffusing though the skin or other membranes before entering the bloodstream.

Drug Concentration

The concentration C of a drug refers to the mass of the drug in a certain volume V of a fluid sample divided by that volume. The mass of the drug is often called the dose D. The expression is written as

$$C = \frac{\text{Mass}}{\text{Volume}} = \frac{D}{V}. \tag{1}$$

The amount of the drug is typically measured in micrograms (μg, or mcg), although milligrams may be used. The fluid sample is usually measured in milliliters. The typical unit for concentration of a drug is μg/mL, or mcg/mL, although it can be expressed as the equivalent unit mg/L.

Fluid samples can be obtained by invasive or noninvasive methods. The primary invasive method requires taking blood from the patient, although other methods include taking samples of spinal fluid or a biopsy of

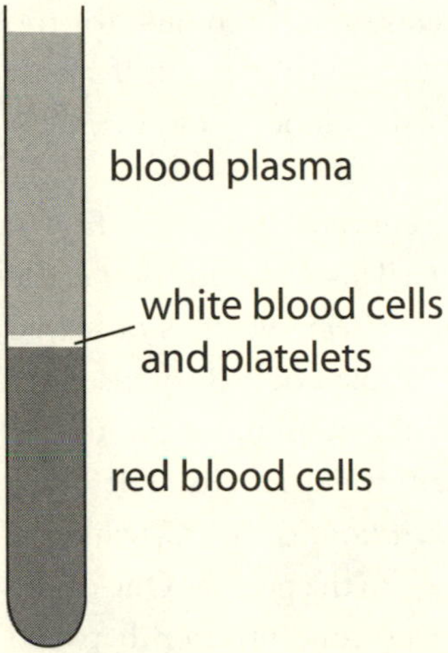

blood plasma

white blood cells
and platelets

red blood cells

FIGURE 8.4. When whole blood is centrifuged it separates into its constituent parts as shown. Drug concentrations are measured in the blood plasma.

tissue. Typical noninvasive methods use urine, although saliva and feces are sampled as well.[1]

The most common method to measure drug concentration requires a sample of blood drawn from the patient. Although blood is a fluid, it is actually composed of a liquid portion and a cellular portion. The liquid component is called plasma and forms a matrix in which blood cells and proteins are suspended. The plasma makes up about 55% of whole blood. The cellular components consist of red blood cells (also called erythrocytes), white blood cells (also called leukocytes), and platelets. The red blood cells make up about 45% of whole blood while the white blood cells and platelets make up less than 1% of whole blood. When whole blood is placed in a centrifuge, the red blood cells separate out at the bottom of the test tube, with the white blood cells and platelets in the middle, and the plasma at the top (fig. 8.4).

For most measurements it is the plasma that is used to determine the drug concentration. Plasma is present in all organs and tissue systems of the body. Thus if a drug is found in a sample of blood drawn from the body, it must be present in the tissues. This statement assumes that the drug concentration in the plasma is in equilibrium with that in the tissues.[2]

If a specific part of the body is the target system for drug treatment, then the concentration of the drug is monitored so as to achieve the desired pharmacological effect, such as a headache that goes away, a tumor that shrinks, or a decrease in congestion. Sometimes the behavior of a person may be an indicator that a drug concentration is too high (or too low). An example is when a person has had too much alcohol. If his or her behavior is suspicious, a measurement of the blood alcohol content helps to confirm the state of the individual.

There are two reference values for drug concentration. The first is the *minimum effective concentration*, or MEC. This concentration is the minimum value that produces a pharmacologic effect, which is observed in the behavior of the person or the response of the body. Responses are observed through visual inspection, physical interaction, or the use of electronic equipment, such as an EKG or brain wave activity. The other reference value is the *minimum toxic concentration*, or MTC, which is the minimum concentration that causes damage to the person. One of the goals of treating a patient is to maintain a drug concentration between these two reference values over a period of time. The region between these two reference values is called the *therapeutic window* or *therapeutic range*. An understanding of the time behavior of drug concentrations in the body is important for the health professional to be able to treat an individual properly.

An example that helps to illustrate the importance of maintaining proper drug concentration has to do with the drug *quinacrine*. Quinacrine is a drug developed during World War II as a substitute for quinine to treat malaria. It was found that small doses of quinacrine were ineffective against malaria. However, larger doses that proved to be effective also turned out to be toxic when they were repeated on a daily basis. Pharmacokinetic studies found that the elimination of quinacrine from the body was quite slow, so that when the drug was given repeatedly its concentration built up over time and exceeded the minimum toxic concentration. To treat the illness effectively, it was determined that large doses should be given initially for a few days to begin fighting the disease, with smaller daily doses given later so that the concentration of quinacrine is maintained within the therapeutic window.[3]

One of the methods used to study drug concentrations is the *plasma level–time graph*. This graph is a plot of the concentration of the drug taken from blood plasma samples as a function of time after a drug has been administered (fig. 8.5). Theoretical curves predict the shape of the graph, which depends on a number of factors, such as the method used to

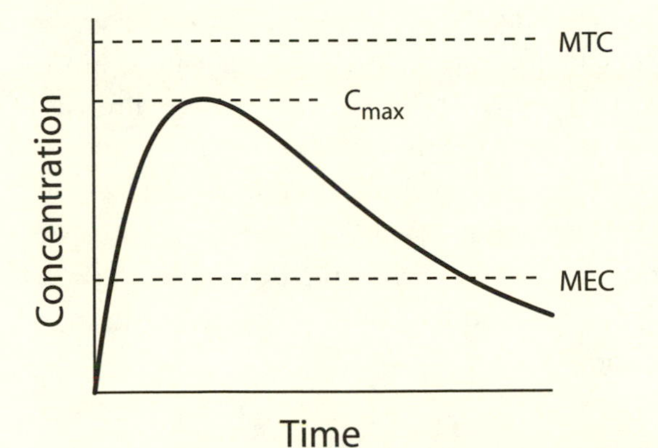

FIGURE 8.5. A plasma level–time graph displays the drug concentration as a function of time. For proper treatment of a patient, the concentration needs to be maintained in the therapeutic window, which ranges from the minimum effective concentration (MEC) to the minimum toxic concentration (MTC). The maximum value of the concentration, C_{max}, should not exceed the MTC.

administer the drug, the frequency of the doses, and the rates of absorption, distribution, and elimination of the drug in the body.

Mathematics

The study of drug concentrations includes a number of mathematical ideas. Concepts such as rate of change, constant values, linear changes, and exponential growth and decay will be considered at various points in the discussion, either separately or collectively depending on the situation. An exponential decay equation similar to that for the decay of parent nuclei as a function of time (see chapter 7) will be considered related to the reduction of the concentration of a drug in the body. In addition, sometimes there will be an exponential growth equation related to the increase of the concentration of the drug. In some cases, growth and decay occur at the same time, so that both have to be considered when describing the time behavior of the concentration.

When a drug enters the body, there are many complex processes taking place. The drug will mix with different fluids and pass through a variety of tissues before and after entering the blood system. If an intravenous injection is given, the drug enters the bloodstream immediately, but is carried to the targeted system. If taken orally, the drug will need to pass through the wall of the GI system to enter the bloodstream. If it is in solid form, it has to dissolve before being absorbed. Once in the circulatory system, it will be distributed throughout the body where it is absorbed in tissues, metabolized, and/or excreted. Each of these processes must be considered in order to fully understand the effect of drug concentration on the response of the patient.

Rates

We have used the term *rate* several times so far. Because drug concentration is dependent on the rate at which a drug enters the system and that at which it leaves the system, we need to define what is meant by rate.

Before proceeding, let me state that we are mainly concerned with the concentration of drug in the circulatory system and not in the tissues. Many of these same concepts of rate apply to the tissues as well. However, because the concentration of drug in the bloodstream is more easily measured and controlled, it can be correlated with the pharmacological effect of the patient much more easily as well.

The term *rate* in this chapter is meant to refer to the rate of change of the concentration as time elapses, or more specifically the change in the concentration ΔC divided by the change in time Δt, or $\Delta C/\Delta t$. Over an extended period of time, the concentration can increase for a while, decrease for a while, and even remain constant. Often, however, we need to monitor dynamic changes in the concentration over much shorter time periods. Thus, we want to consider time intervals Δt that are very small. Those who have studied calculus will likely recognize that the rate for very small time intervals is the *derivative* of the concentration with respect to time, or dC/dt. I will continue to use the delta symbol Δ instead of the derivative symbol d in the expression for rate.

The overall rate of change of concentration depends on the three processes: absorption, distribution, and elimination. All of these processes have their own individual rates that contribute to the overall rate. Mathematically, we write the rate of change in the drug concentration as the sum of these three rates, or

$$\begin{matrix}\text{Rate of Change in} \\ \text{drug concentration}\end{matrix} = \begin{matrix}\text{Absorption} \\ \text{rate}\end{matrix} + \begin{matrix}\text{Distribution} \\ \text{rate}\end{matrix} + \begin{matrix}\text{Elimination} \\ \text{rate}\end{matrix}.$$

In symbol form, we can express the rate equation as[4]

$$\frac{\Delta C}{\Delta t} = \left(\frac{\Delta C}{\Delta t}\right)_a + \left(\frac{\Delta C}{\Delta t}\right)_d + \left(\frac{\Delta C}{\Delta t}\right)_e. \qquad (2)$$

During the absorption process, drug enters the bloodstream; thus, the absorption rate is positive. During distribution and elimination, drug leaves the bloodstream; thus, the distribution rate and the elimination rate are both negative.

The overall rate is determined by the relative sizes of these three individual rates. There are three distinct situations that occur. First, if the overall concentration increases, the rate is positive. This is the case when the absorption rate is greater than the sum of the distribution and elimination rates. Second, if the overall concentration decreases, the rate is negative. Here, drug is removed from the system during the distribution and/or the elimination process at a faster rate than it is being absorbed. Third, if the concentration remains constant, the overall rate is zero because $\Delta C = 0$. In this situation the absorption rate is equal to the sum of the distribution and elimination rates. When the overall rate is zero the concentration is unchanged over time.

Order of Reaction

All movements of drugs into and out of the bloodstream have rates that follow fairly specific mathematical trends. Typically, the rates fall into two categories. The first is where the rate is constant, called a *zero-order* reaction or process, and the second is where the rate is proportional to the concentration, called a *first-order* reaction or process. Higher-order reactions are possible, but most of the experimental drug concentration measurements can be predicted using a zero-order process, a first-order process, or a combination of the two.

Zero-Order Reaction

In a zero-order reaction, the rate of change of the drug concentration is constant and does not depend on the actual concentration of the drug. This means that the amount of drug that enters or leaves the bloodstream changes by the same amount over a constant time interval. It can be expressed mathematically as

$$\frac{\Delta C}{\Delta t} = \pm k_0, \tag{3}$$

where k_0 is a constant value called the zero-order rate constant and has units of concentration/time. The rate is positive if the concentration of the drug increases with time, negative if it decreases with time, and zero if it is constant.

In calculus, the expression can be written as a differential equation, which can be solved for the drug concentration as a function of time:

$$C(t) = \pm k_0 t + C_0, \tag{4}$$

where $C(t)$ is the concentration at time t, k_0 is the zero-order rate constant, and C_0 is the initial concentration at time $t = 0$. Equation (4) is simply the equation of a straight line where $\pm k_0$ is the slope of the line and C_0 is the y-intercept. Thus, the rate in equation (3) corresponds to the slope of a straight line graph.

Figure 8.6 shows graphs of concentration versus time for three possibilities of the slope—zero, positive, and negative. The graph with zero slope (fig. 8.6a) is described mathematically with the equation $C(t) = C_0$. It shows that the concentration remains constant over an extended period of time. This situation occurs when a patient receives an IV bolus or injection where there is no elimination. The graph with the positive slope (fig. 8.6b) is described mathematically using the equation $C(t) = k_0 t + C_0$. It shows that the concentration increases linearly with time. We will see later that this situation can occur when an IV infusion is given to a patient where, once again, there is no elimination. The graph with the negative slope (fig. 8.6c) is described mathematically with the equation $C(t) = -k_0 t + C_0$. It shows that the concentration decreases linearly with time. There are only a few cases that exhibit such zero-order elimination,

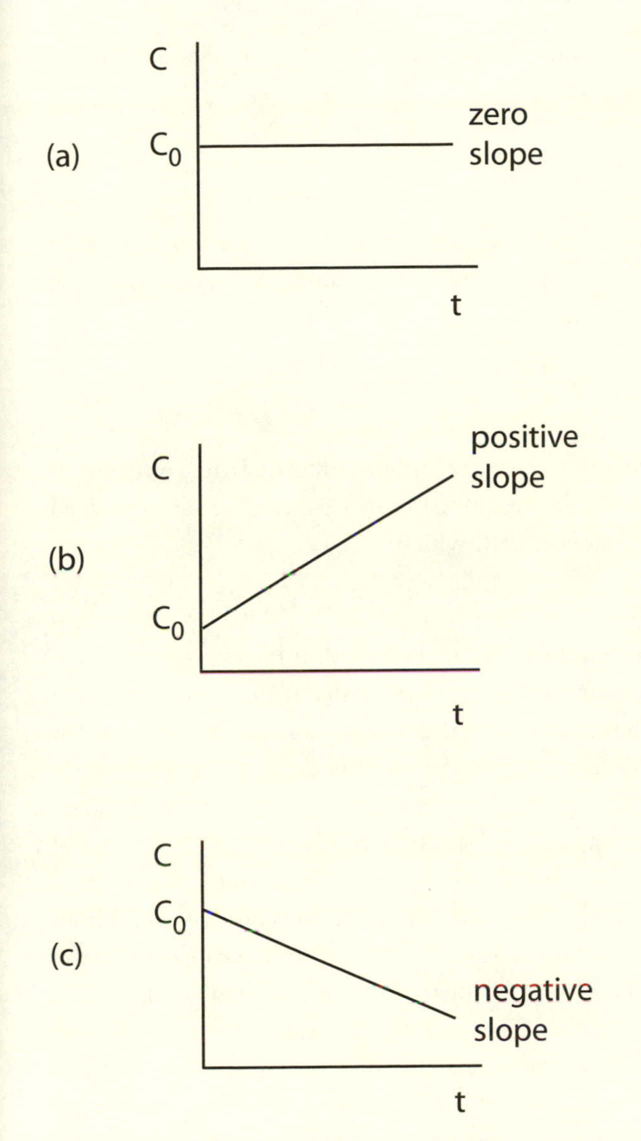

FIGURE 8.6. Zero-order reactions result in linear concentration functions C with time t starting with an initial concentration of C_0. (a) If the slope is zero, then the concentration is constant. (b) If the slope is positive, the concentration increases linearly with time. (c) If the slope is negative, the concentration decreases.

in particular alcohol and salicylates. For most drugs, however, elimination is a first-order reaction.

First-Order Reaction

A first-order reaction is where the rate of change of the concentration of drug in the system is proportional to the actual concentration. It is expressed mathematically as

$$\frac{\Delta C}{\Delta t} = \pm kC, \qquad (5)$$

where k is the first-order rate constant and has units of 1/time or $(\text{time})^{-1}$. The differential equation that results from this expression can be solved for the concentration function $C(t)$, which is

$$C(t) = C_0 e^{\pm kt}, \qquad (6)$$

where C_0 is the initial concentration at time $t = 0$. If the exponent is positive, then the concentration grows exponentially large. If the exponent is negative, then we have exponential decay similar to the results from chapter 7 on nuclear decay; k is called the decay constant.

Figure 8.7 shows two graphs, one for exponential growth with $C(t) = C_0 e^{+kt}$ and the other for exponential decay with $C(t) = C_0 e^{-kt}$. Exponential growth, as illustrated in the first graph (fig. 8.7a), does not actually occur in the body, so it appears that first-order absorption is not possible. However, it does occur when oral medication is taken, and because this is the most common method of drug administration, we will explore first-order absorption in more detail later. The second graph (fig. 7.8b) for exponential decay shows the concentration decreasing with time. Drug distribution in the body and drug elimination from the body follow first-order reactions, and the concentration decreases according to the exponential decay expression.

Compartment Models

The overall rate of change of the drug concentration in the body is based on the rates of drug absorption, distribution, and elimination. But each of

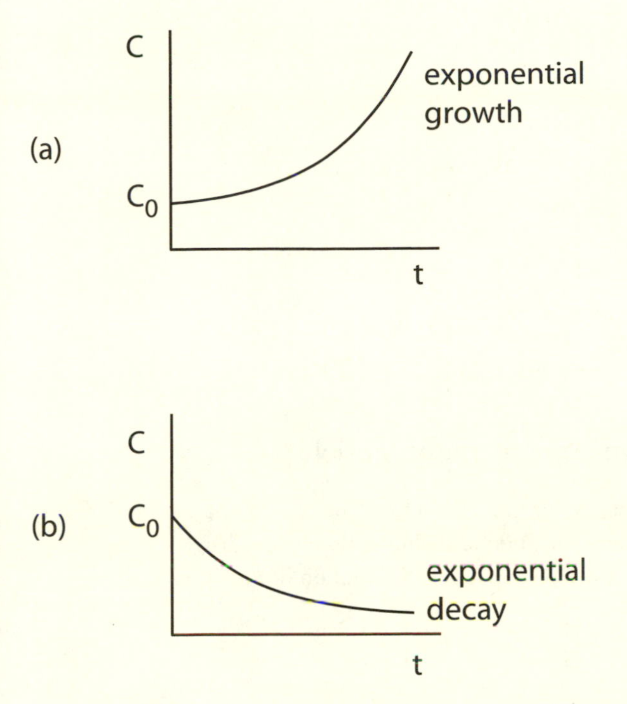

FIGURE 8.7. First-order reactions result in exponential concentration functions C with time t starting with an initial concentration of C_0. (a) For exponential growth, the concentration increases with time. (b) For exponential decay, the concentration decreases to zero asymptotically.

these rates contributes in a variety of ways with different slopes and rate constants. The mathematics can get quite complicated.

To take these processes and other considerations into account, mathematical models are used to determine theoretical formulas so that drug concentration can be predicted over the course of treatment. These models are used to predict drug concentrations depending on the type and frequency of dosing, to determine proper dosage procedures, to estimate possible accumulation of drugs or their products, to relate the concentration of the drug to pharmacological effects, to help determine differences in different forms of the drug, to describe how changes in the progress of treatment affects the use of the drug in the body, and to investigate interactions with other drugs.[5] Let's look more closely at two of these models.

The first model we consider is called the *one-compartment model.*

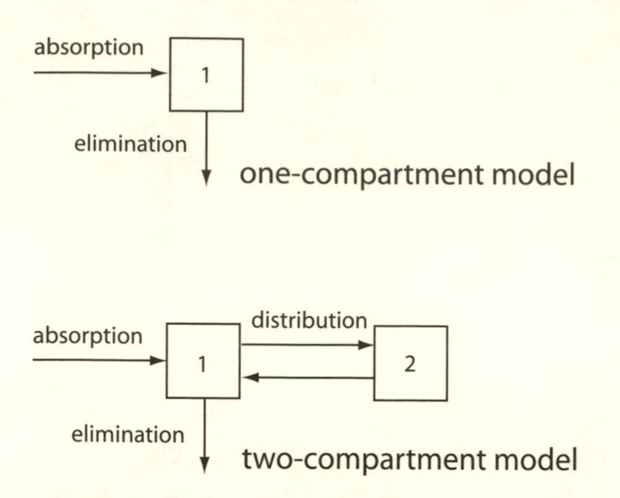

FIGURE 8.8. One- and two-compartment models. The one-compartment model takes into account absorption and elimination of the drug, whereas the two-compartment model also includes distribution of the drug between compartments.

In this model we consider the blood system to be similar to a tank containing a fluid. The tank is open, which means that drug can be added and removed (fig. 8.8). As drug is added or removed, the total volume of fluid remains constant over time. This model is similar to that for the circulatory system in the human body where the volume remains (almost) constant with time.

What happens if an amount of drug is added to or removed from the tank? If a drug is added to the tank and is completely mixed with the fluid rapidly (the system is "well stirred"), then the concentration of the drug will depend on the mass of the drug in the tank divided by the volume of the tank. As drug is added to and removed from the tank, the concentration of the drug in the tank may change over time depending on the rates of addition and removal.

Drug can enter the human bloodstream by one of several methods of administration; this is the absorption process. Drug leaves the bloodstream as it is absorbed by tissues, metabolized in a chemical reaction, or excreted through waste material. In the one-compartment model, these three ways of leaving the bloodstream are referred to collectively as the elimination process. (Thus, there is no separate rate dealing with the distribution process.) If the rate at which medication is added to the system is the same as the rate at which it leaves the system, the concentration remains constant; if the rates are different, the concentration varies. In a bit, we will explore the situations where the drug enters the bloodstream quickly by an IV bolus or injection, at a steady rate by an IV infusion, and at a variable rate by oral medication.

The next model to consider is called the *two-compartment model.* In this model, the blood system is considered the primary tank or compartment where the drug enters the body, and the various tissues where the drug is supposed to go during the distribution process are considered to be a second tank (fig. 8.8). As with the one-compartment model, the two compartments are open. Thus, drug can enter and leave both the first and the second compartments. Medication is added to the first compartment as in the one-compartment model, but moves out of the first compartment to the second compartment (distribution) and/or to other parts of the body for elimination. The second compartment has drug moving into it from the first compartment, but an important point about the two-compartment model is that *medication moving out of the second compartment can go back into the first compartment.*

The concentration of the drug in the two compartments varies with time according to the rate at which the drug enters and leaves the compartments. For the first compartment, the main rate at which the drug enters the compartment depends on the method of administration. There are two rates for drug leaving the compartment: one for drug going to the second compartment during the distribution process and another for drug being eliminated through either metabolic processes or excretion. However, there is another consideration: the rate for drug entering the first compartment coming from the second compartment. All four of these rates (two entering and two leaving) are needed to determine the drug concentration in the first compartment as a function of time. In this chapter, we will not consider the time-dependent concentration in the second compartment, although it may be needed by the scientist or pharmacist in order to link the pharmacokinetics to the pharmacological effect.

The rate at which the drug flows out of the first compartment to the second compartment is different from the rate of flow from compartment

2 to compartment 1. If the rates were the same, then the concentration of compartment 1 would exhibit the same time-dependent behavior as in the one-compartment model. As in the one-compartment model, a drug can enter the first compartment quickly (IV injection), at a steady rate (IV infusion), or orally.

Perhaps you can see that there are other models that are much more complicated. Models that have more than two compartments can have the compartments arranged in series or parallel. A series arrangement is one where the compartments are connected one after another like boxcars of a train. Each compartment exchanges drug with each neighbor on either side. In contrast, a parallel arrangement is one where the central compartment (the circulatory system) is connected with each peripheral compartment (various tissue systems and organs) and drug is exchanged with each one simultaneously and at different rates. It is important for the physician or pharmacist to consider these more complicated models when monitoring a dosing regimen. However, we will look mainly at the one-compartment model because it is a much simpler problem to solve than the two-compartment model. In addition, the one-compartment model is very accurate for many real situations.

Reactions and Compartment Models

Recall that the overall rate of change of the concentration of drug in the body depends on the absorption rate, the distribution rate, and the elimination rate. Theoretically, the rates follow a zero-order or a first-order process. In practice, however, the rates follow certain behaviors depending on the dosing method and compartment model.

In the one-compartment model, there are only two terms in the rate equation: the absorption and the elimination rates. The absorption rate is zero order or first order depending on the dosing method. For an IV bolus, the absorption (or availability) is instantaneous, which means that the concentration goes to a certain value immediately; it is a zero-order reaction where $\Delta C/\Delta t = 0$. For an IV infusion, the rate of absorption (or availability) is constant and depends on the rate of infusion; it is a zero-order reaction as well where $\Delta C/\Delta t = +k_0$. For oral medication and other forms that do not go directly into the bloodstream, the absorption follows a first-order reaction where $\Delta C/\Delta t = +kC$. The elimination rate is usually a first-order reaction where $\Delta C/\Delta t = -kC$.

In the two-compartment model, there are three terms in the rate equation: the absorption rate, the distribution rate, and the elimination rate. The absorption and elimination rates follow the same order reactions as in the one-compartment model. However, the distribution rate from the main compartment to the other compartments is a first-order reaction. In addition, recall that drug can flow back into the first compartment from the second compartment; this rate is a first-order reaction as well.

Remember that all first-order reactions have a rate constant k. Thus, there is a rate constant for absorption, k_a, a rate constant for elimination, k_e (which includes metabolism and excretion), and rate constants for exchanges between compartments (which includes distribution). For example, for drug moving from compartment 1 to compartment 2 the rate constant may be written k_{12}, but there is a different rate constant for drug moving from compartment 2 to compartment 1 given by k_{21}.

The multicompartment models can be quite complicated. Each compartment "communicates" with others as a first-order reaction. From now on, however, we will discuss almost exclusively the one-compartment model.

One-Compartment Model

For the one-compartment model, the rate of change in the drug concentration is a combination of the absorption rate and the elimination rate, or

$$\frac{\text{Rate of Change in}}{\text{drug concentration}} = \frac{\text{Absorption}}{\text{rate}} + \frac{\text{Elimination}}{\text{rate}}.$$

There is no distribution term to concern ourselves with here although it comes up in the two-compartment model. The absorption rate depends on the method of drug administration, and also on other factors, such as the dosage form, the solubility, movement through the GI wall, and so on, but the same mathematical concepts apply to all of these other factors for first-order absorption. The elimination rate is first order, so it depends on the concentration of the drug in the system regardless of the method of administration. In mathematical symbols, we can express the rate equation as[6]

$$\frac{\Delta C}{\Delta t} = \left(\frac{\Delta C}{\Delta t}\right)_a + \left(\frac{\Delta C}{\Delta t}\right)_e . \qquad (7)$$

This equation can be solved to yield an expression that gives the drug concentration as a function of time, written as $C(t)$. A plot of this function gives us a plasma level–time graph (described earlier).

As we explore in more depth the one-compartment model, we will look at various combinations of the absorption and elimination terms in equation (7), and we will describe how the results differ based on whether the drugs are administered as an IV bolus or injection, an IV infusion, or orally.

Zero-Order Absorption with No Elimination: IV Bolus

The graph in figure 8.6a with slope of zero is an example of zero-order absorption for an IV bolus when there is no elimination. This graph is described mathematically with the equation $C(t) = C_0$. The medication goes into the bloodstream immediately and is mixed throughout the body very quickly. The concentration C_0 (determined from the dose D_0 and the volume of distribution V) remains constant over time when there is no elimination. This situation does not actually occur in practice, but we will include elimination later.

If multiple doses are given, the concentration increases in a stepwise manner. When the second dose is given, the concentration jumps to a higher value almost immediately and remains at that value. When another dose is given, it jumps up to an even higher value and so on. Figure 8.9 shows the graph that is expected for multiple doses by IV bolus with no elimination.

Zero-Order Absorption with No Elimination: IV Infusion

Figure 8.6b with positive slope is an example of zero-order absorption for IV infusion when there is no elimination. The graph is described mathematically using the equation $C(t) = k_0 t + C_0$. It shows that the concentration increases linearly with time. The drug enters the patient at a constant

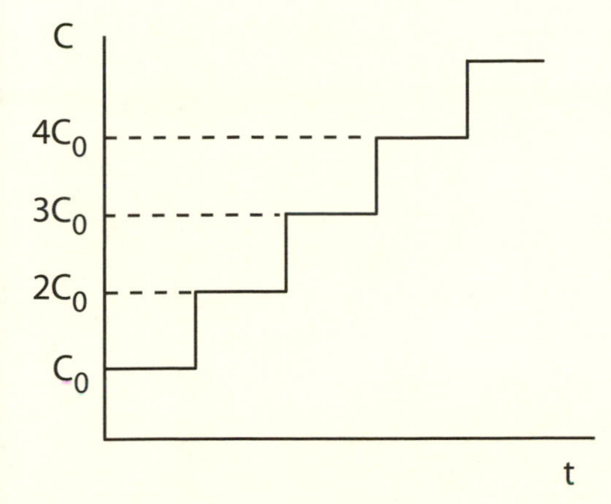

FIGURE 8.9. When multiple IV bolus doses are given, the concentration increases each time by an amount equal to C_0. If there is no elimination of the drug from the body, the concentration remains constant until the next dose is given, increasing to a new concentration with each injection.

rate k_0, called the infusion rate, which is controlled by hanging the IV bag high enough with a drip-rate controller or flow-rate monitor attached in line with the tube that goes from the bag to the needle. If there is no drug in the system when the IV is started, then $C_0 = 0$. The concentration increases until the bag is empty or the infusion is stopped, and then the concentration remains at a constant value (fig. 8.10a).

If multiple IV infusions are given, the initial concentration will not be zero after the first IV bag is emptied. If the IV infusion occurs on a continual basis, with one IV bag immediately replaced with another bag and so on, the concentration continues to build up linearly (fig. 8.10b). If the IV bag empties and is not replaced for a while, the concentration remains at a constant level until another infusion begins. At that time, the concentration increases linearly again, and the process continues. Figure 8.10c illustrates the concentration as a function of time for this situation.

No Absorption and First-Order Elimination

In first-order elimination, the concentration of the drug in the bloodstream decreases with time exponentially, following the equation

$$C(t) = C_0 e^{-k_e t}, \qquad (8)$$

where k_e is the elimination rate constant (see fig. 8.7b). Because the rate of decrease is proportional to the amount of drug present, the concentration decreases quickly. As time elapses, however, the rate of decrease slows

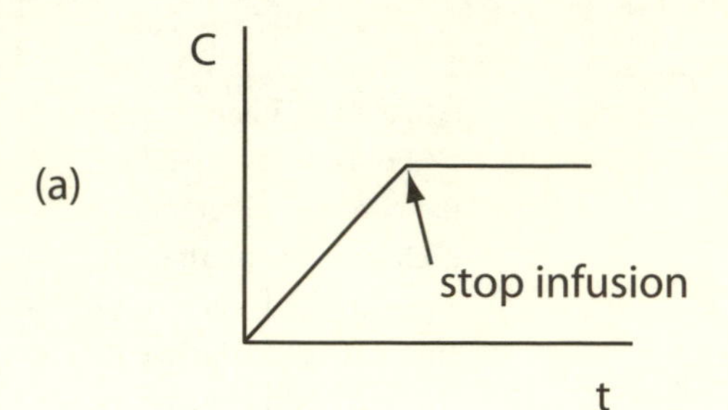

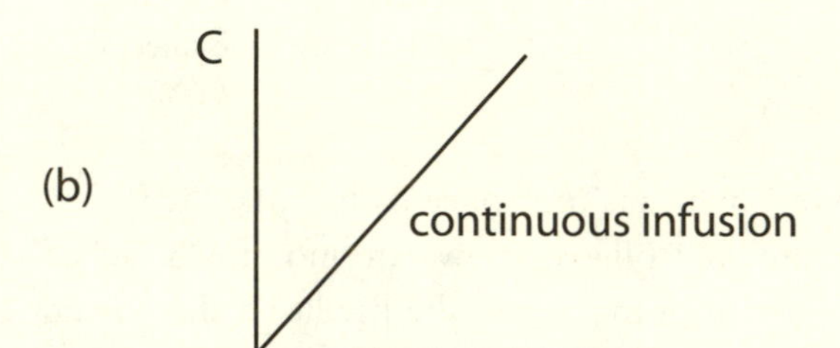

C
(c)
several infusions
t

FIGURE 8.10. When an IV infusion is given, the concentration of the drug increases linearly if there is no elimination. (a) The infusion is stopped at a certain time. (b) The infusion continues with an empty bag immediately replaced by another, and so on. (c) The infusion is stopped for a time, started again, stopped, started again, and so on.

down because the concentration is smaller, and eventually it gets closer and closer to zero.

Pharmacists often need to determine the decay constant for a particular drug. So rather than plotting concentration versus time, they plot the natural logarithm of the concentration versus time. If we take the natural logarithm of equation (8), we get

$$\ln C = -k_e t + \ln C_0. \quad (9)$$

A plot of ln C versus t gives a straight line graph with a slope of $-k_e$ (fig. 8.11).

The half-life $t_{1/2}$ for a drug in the bloodstream is defined as the time that it takes for half of the drug to be removed from the system. This definition is almost identical to the half-life used in chapter 7 for nuclear radiation. It can be shown that the half-life for a first-order elimination process is

$$t_{1/2} = \frac{\ln 2}{k_e}.$$

It turns out that the half-life for a first-order process is constant. This means that the time it takes for the concentration to decrease from its initial value C_0 to half this value $C_0/2$ is the same as the time it takes to decrease from $C_0/2$ to $C_0/4$. Another period of time equal to one half-life elapses when the concentration decrease from $C_0/4$ to $C_0/8$, and so forth. For a zero-order process, the half-life is not constant, so it is not used.

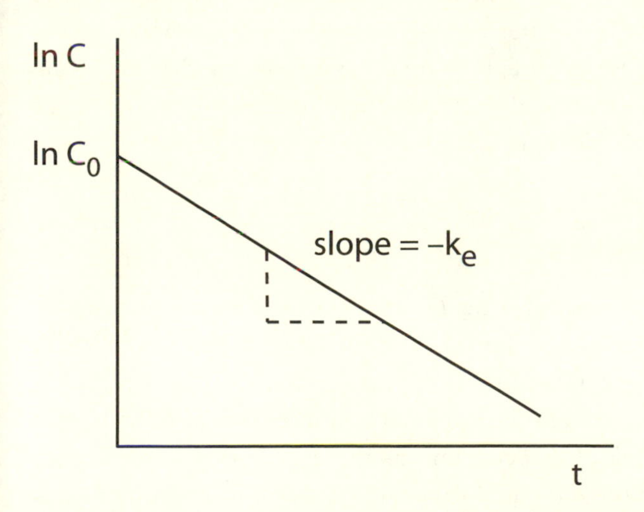

FIGURE 8.11. For first-order elimination with no absorption, the concentration decreases exponentially. The natural logarithm of the concentration function is a straight line with a negative slope given by the elimination constant k_e.

First-Order Absorption with no Elimination: Oral Medication

As stated previously, exponential growth is not realistic for drug concentrations in the body (or in any other system for that matter), so it would appear that first-order processes do not apply when drug is absorbed in the body. When medication is given by IV bolus or IV injection, these are zero-order processes where absorption is instantaneous and the drug concentration reaches a starting value almost immediately for IV bolus or increases linearly based on a constant drip rate for IV infusion.

When medication is taken orally, a first-order process occurs. Exponential growth assumes that there is an unlimited supply of medication readily available, but in an oral dose there is only a finite amount of medication available. As the oral medication dissolves or is absorbed in the GI wall, the amount available to pass into the bloodstream decreases exponentially according to a first-order process. Thus, there is a combination of a first-order process resulting in exponential decay of medicine available from the oral dose and a first-order process as the medication enters the bloodstream.

The rate equation for this process shows that the changing concentration of the drug entering the blood stream, $\Delta C/\Delta t$, is proportional to the concentration available to enter the bloodstream from the oral medication, $C_0 e^{-k_a t}$, which decreases exponentially. (We assume that all of the drug in the oral dose will be absorbed in the bloodstream; this does not always happen.) It is written as

$$\frac{\Delta C}{\Delta t} = k_a C_0 e^{-k_a t}, \qquad (10)$$

where C_0 is the total dose of drug D_0 available in the oral medication divided by the volume of distribution V in the bloodstream where it will end up. The constant k_a is the first-order rate constant as the medication dissolves and passes through the stomach and intestines into the bloodstream. You can think of this constant as the distribution rate constant for the oral medication; it is equal to the absorption rate constant as it is absorbed into the bloodstream.

The differential equation (10) yields the concentration function

$$C(t) = C_0(1 - e^{-k_a t}). \qquad (11)$$

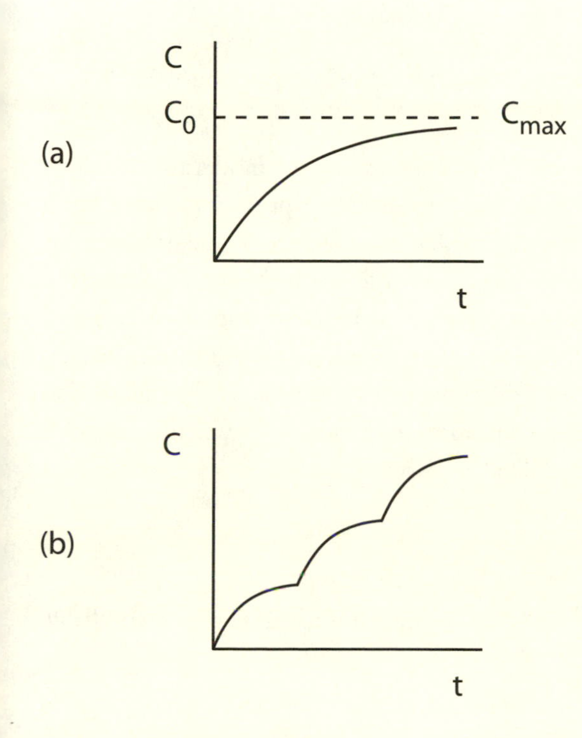

FIGURE 8.12. First-order absorption of an oral dose with no elimination. (a) The concentration approaches a maximum value C_{max} that is equal to the amount of drug in the original dose C_0. The concentration remains constant at the maximum value. (b) Multiple doses of oral medication cause the maximum concentration to increase with each dose taken.

A plot of equation (11) is shown is figure 8.12a. Notice that the concentration increases quite rapidly at first, but then more slowly approaches the maximum concentration C_{max}, which is just the expected concentration $C_0 = D_0/V$. The horizontal line at C_0 is called an asymptote, and the concentration approaches the value of C_0 asymptotically. If no other dose is given, the concentration remains constant at C_0 as long as there is no elimination.

If multiple doses are given, the concentration increases in a manner that is not linear, nor strictly stepwise. When a second dose is given, the concentration increases asymptotically to a new value for C_0. When another dose is given, it increases again to a higher value, and so on (fig. 8.12b). Pharmacokinetics studies have found that not all the medication in the oral dose is absorbed by the bloodstream; some of it is excreted through waste without being absorbed. In this situation, equation (10) is modified to include a multiplier F, which is the fraction of the dose absorbed. For our discussion, we will assume that all of the medication is absorbed.

Absorption with Elimination in the One-Compartment Model

We are now ready to explore the one-compartment model for the three routes of administration where we consider absorption and elimination together. The one-compartment model is illustrated in figure 8.13 for the IV bolus, the IV infusion, and the oral dose. In the figure, the arrow pointing into the compartment represents the absorption process, and the arrow pointing out of the compartment represents the elimination process. Mathematically, we will represent each type of administration as a variation of the main rate equation [equation (7)], which is repeated here:

$$\frac{\Delta C}{\Delta t} = \left(\frac{\Delta C}{\Delta t}\right)_a + \left(\frac{\Delta C}{\Delta t}\right)_e . \qquad (12)$$

We will look at what happens when a single dose and then when multiple doses are given.

IV Bolus with Elimination

When a drug is administered using an IV bolus, a certain dose (or mass) of drug is injected into the bloodstream. It is assumed that the drug mixes immediately in the circulatory system; this is not really true of course,

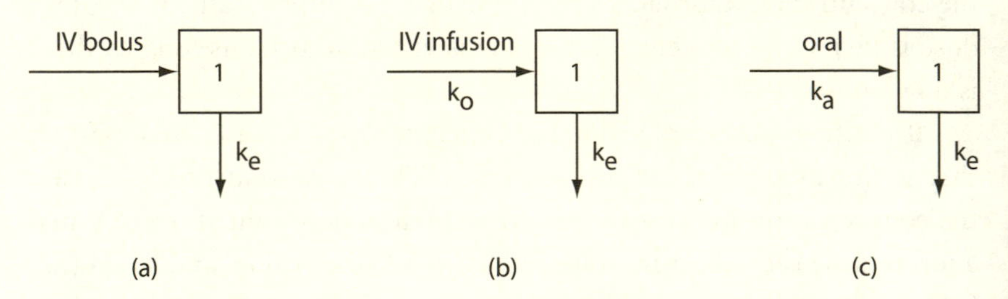

FIGURE 8.13. Three different examples of medication entering the one-compartment model (absorption) with drug leaving the compartment at the elimination rate constant k_e. (a) For an IV bolus, the dose is injected all at once. (b) For IV infusion, the drug enters at the infusion rate k_0. (c) For oral medication, the drug enters at the absorption rate constant k_a.

but compared to the time scales required for elimination, the time for the drug to mix in the bloodstream is very short.

When an injection is given, not only does absorption begin immediately, but elimination begins immediately as well. As discussed previously, if there is *no elimination* of drug in the body, then the concentration remains constant and the absorption rate in equation (12) is zero, or

$$\left(\frac{\Delta C}{\Delta t}\right)_a = 0. \tag{13}$$

Even with elimination present, the absorption rate is zero as well. Because the elimination rate is first order and depends on the concentration, then the rate of elimination is

$$\left(\frac{\Delta C}{\Delta t}\right)_e = -k_e C, \tag{14}$$

where k_e is the elimination rate constant with units of 1/time. Substitution of equations (13) and (14) into equation (12) gives an overall rate equation of the concentration for the IV bolus dosing method of

$$\frac{\Delta C}{\Delta t} = -k_e C. \tag{15}$$

This differential equation can be solved for $C(t)$ to yield

$$C(t) = C_0 e^{-k_e t} = \frac{D_0}{V} e^{-k_e t}. \tag{16}$$

Suppose a patient is given a dose of a drug such that $D_0 = 100$ mg. Also suppose that the volume into which the drug is given (called the volume of distribution) is $V = 2$ L. Finally, suppose that the elimination decay constant for this drug is $k_e = 0.277$/h. These numbers give us an initial concentration $C_0 = 50$ μg/mL, and the expression becomes

$$C(t) = 50\, e^{-0.277t}. \tag{17}$$

In figure 8.14 we plot the concentration function versus time. The half-life can be calculated as $t_{1/2} = \ln 2/k_e = 2.5$ h, which means it takes 2.5 h for half of the drug to be eliminated from the body, either through meta-

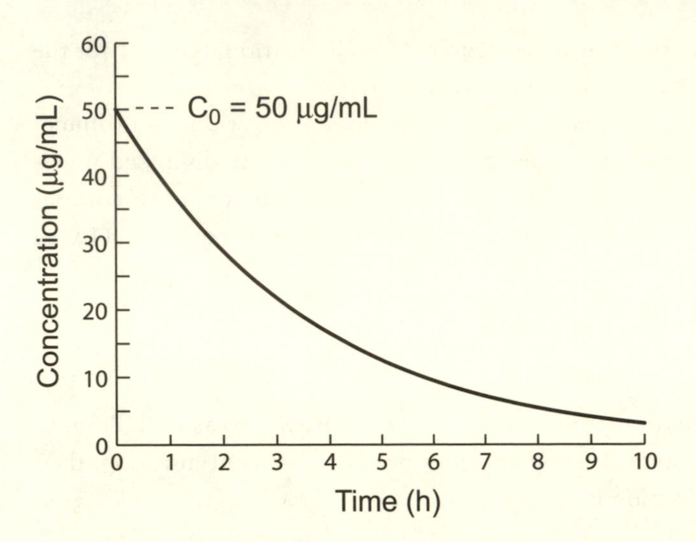

FIGURE 8.14. Concentration versus time for an IV bolus dose with elimination. The initial concentration of the medication is $C_0 = 50$ μg/mL.

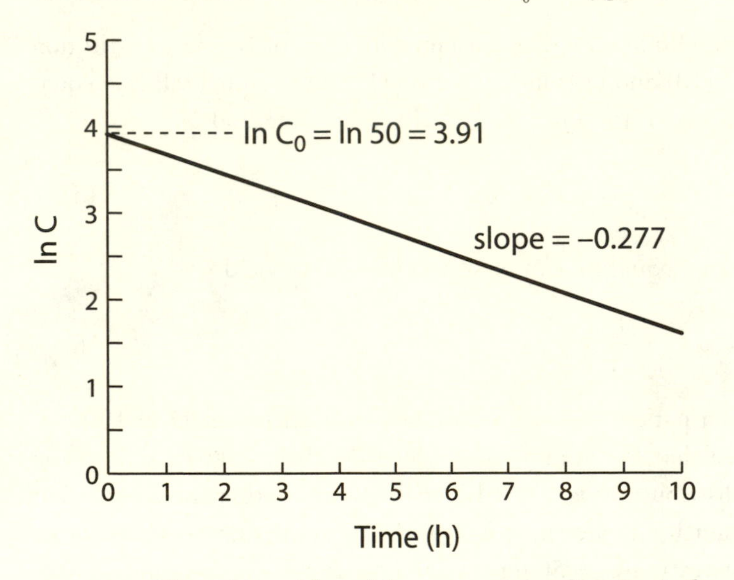

FIGURE 8.15. Natural logarithm of the concentration function in figure 8.14.

bolic processes or through excretion. In figure 8.15, we plot the natural logarithm of the concentration as a function of time, which is a straight line graph with a slope of $k_e = -0.227$ h^{-1}. It crosses the vertical axis at the value ln 50 (the y intercept). The equation of this line is

$$\ln C(t) = \ln 50 - 0.277t. \quad (18)$$

Thus we see that in the IV bolus method of administration the concentration starts at its initial value and decays exponentially. This method and the mathematics that result are the simplest example of predicting drug concentrations in the body.

Multiple IV Bolus Doses

Sometimes it is necessary to give multiple doses rather than just one dose, to maintain a concentration high enough to produce the desired pharmacological effect. The mathematics is almost identical to that for the single-dose exponential decay process that we have just examined. But the main consideration is that, if there is any drug still in the bloodstream when another dose is given, the additional drug is simply added to what was left from the previous injection.

Consider the example above that resulted in equation (17). Suppose that four hours after the first injection another injection identical to the first one is given. The concentration of the first dose will be about 33% of the starting concentration at the end of four hours:

$$C(t) = 50\, e^{-0.277t},$$
$$C(4) = 50\, e^{-0.277(4)} = 50 \times 0.33$$
$$= 16.5\ \mu\text{g/mL}.$$

When the second injection is given, the concentration jumps immediately by 50 μg/mL to a new starting concentration of 66.5 μg/mL, which then decays exponentially. If another dose is given four hours later when the concentration has decreased to about 22 μg/mL (33% of 66.5), the concentration jumps to about 72 μg/mL. If many repeated doses are given at regular four-hour intervals, the concentration can be represented by the graph shown in figure 8.16.

Notice that for repeated doses the concentration has a maximum value C_{max} and a minimum value C_{min} that levels out after several doses. These are called the *steady-state* values because they do not change if the dosing regimen continues. For proper treatment of a patient, one of the goals is to maintain C_{max} and C_{min} within the therapeutic window, that is, greater than the minimum effective concentration (MEC) and smaller

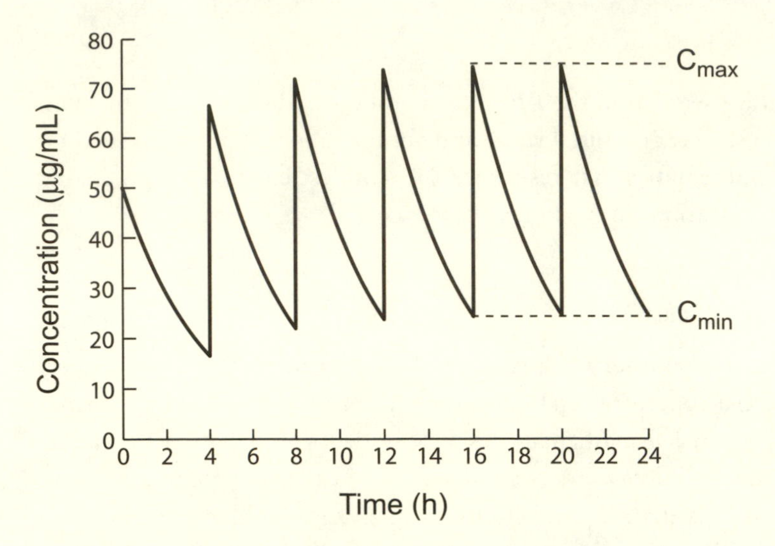

FIGURE 8.16. Multiple IV bolus doses given every four hours show an immediate increase in concentration followed by exponential decay during elimination. Steady-state values for C_{max} and C_{min} are reached after about five doses.

than the minimum toxic level (MTC). This maintenance of the concentration depends on the amount of medication in the injection and how often the medication is given. Thus, you can see the importance of knowing the pharmacokinetic behavior of a drug.

IV Infusion with Elimination

A drug administered using IV infusion enters the bloodstream at a specific rate. The drug mixes with blood in the circulatory system (almost) immediately, and elimination also begins immediately. In the overall rate equation (12), the absorption rate is equal to the infusion rate k_0, which has units of concentration per time. Thus,

$$\left(\frac{\Delta C}{\Delta t}\right)_a = k_0. \tag{19}$$

The elimination rate is first order (just as it is for the IV bolus), so it is written

$$\left(\frac{\Delta C}{\Delta t}\right)_e = -k_e C, \tag{20}$$

where k_e is the elimination rate constant with units of 1/time. Substitution of equations (19) and (20) into equation (12) gives an overall rate equation expressed as

$$\frac{\Delta C}{\Delta t} = k_0 - k_e C. \quad (21)$$

This differential equation can be solved for $C(t)$ to yield

$$C(t) = \frac{k_0}{k_e}(1 - e^{-k_e t}) = \frac{D_r}{Vk_e}(1 - e^{-k_e t}), \quad (22)$$

where D_r is called the dose rate and has units of mass per time.

Consider a patient receiving an infusion with a dosing rate of 15 mg/h into a volume $V = 5$ L; take the elimination half-life of the drug in the patient to be 3 h. These values give us an infusion rate $k_0 = 3$ μg/mL/h, and an elimination decay constant of $k_e = \ln(2)/t_{1/2} = \ln(2)/3\text{ h} = 0.231\text{ h}^{-1}$. Equation (22) becomes

$$C(t) = 13.0\ \mu\text{g/mL}\ (1 - e^{-0.231t}). \quad (23)$$

In figure 8.17 we plot the concentration function versus time, showing the concentration approaching the maximum value $C_{max} = 13.0$ μg/mL.

Recall that, when IV infusion occurs with no elimination, the concentration increases linearly with no upper bound if the infusion continues for a long time. When elimination is present, the concentration approaches the maximum value C_{max} after some period of time and remains at that level as long as the infusion continues. This maximum value is called the steady-state concentration. If the infusion were to be stopped, the drug concentration would decay exponentially. An example is shown in figure 8.18 where the infusion in the previous example is stopped after 10 h. Thus there is a combination of a rise in concentration exponentially approaching a maximum value when the infusion starts, and an exponential decay when the infusion is stopped. If multiple infusions are given, the growth and decay continue, but the maximum concentration will not exceed the value of C_{max} that would occur for continuous infusion.

Infusions are often given to maintain a certain drug concentration for a long period of time. As was the case with the IV bolus, the goal is to maintain a concentration of drug in the system that provides the desired pharmacological effect. It turns out that the maximum concentration de-

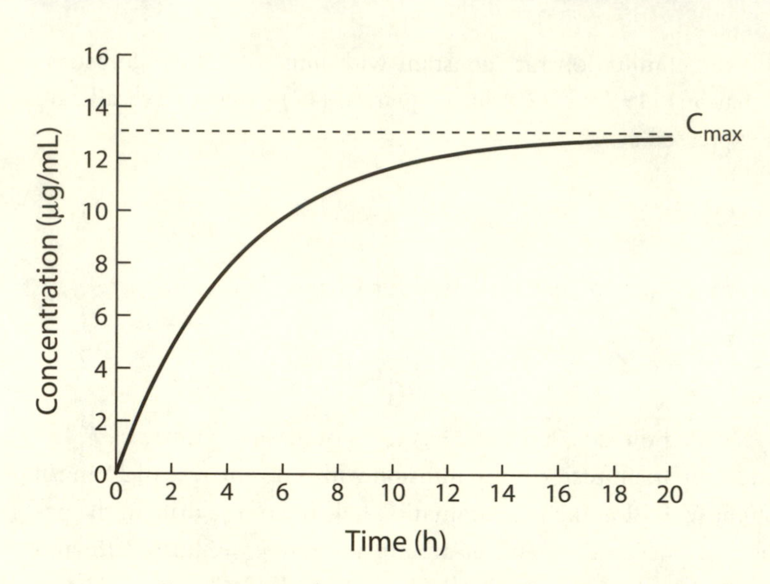

FIGURE 8.17. Concentration function versus time for IV infusion. In this example the concentration function for the IV infusion approaches a maximum value of 13.0 μg/mL.

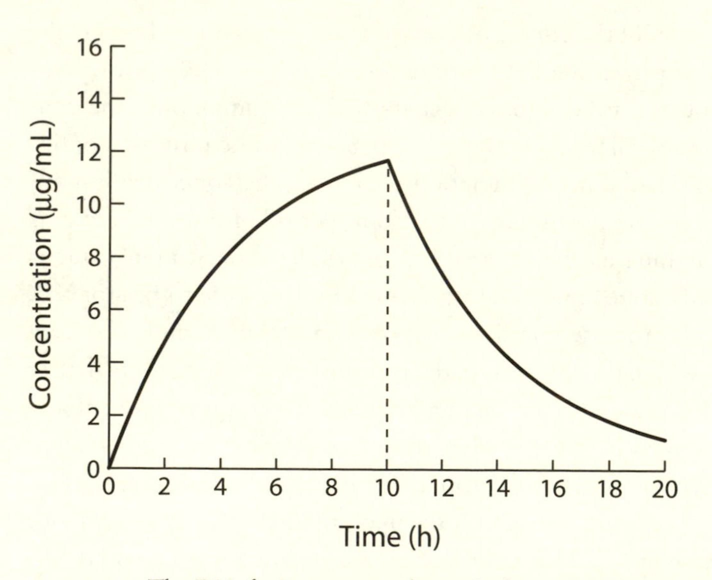

FIGURE 8.18. The IV infusion process shown in figure 8.17 is stopped after 10 h, resulting in exponential decay of the concentration.

pends on both the infusion rate k_0 and the elimination rate constant k_e given by $C_{max} = k_0/k_e$ To adjust the maximum concentration for proper treatment of the patient the infusion rate must be set properly. If the infusion rate is too high, then the maximum concentration may exceed the minimum toxic concentration. If the infusion rate is too small, the maximum concentration may be below the minimum effective concentration.

IV Infusion plus IV Bolus Dose

There are situations where the desired steady-state concentration for IV infusion needs to be obtained rapidly. To accomplish this, an IV bolus dose is given initially. If you think about it, an IV bolus is simply an IV infusion with a very large infusion rate because the entire dose is given at once. If the proper amount of medication is given in the injection followed by the proper amount of medication given more slowly by infusion, the concentration can be maintained at a constant value from the very start of the process. The initial IV bolus dose is called a *loading dose*.

To see how this works, let's consider the concentration function for the IV bolus dose given by

$$C(t) = \frac{D_0}{V} e^{-k_e t} \tag{24}$$

and the concentration function for IV infusion expressed as

$$C(t) = \frac{D_r}{Vk_e}(1 - e^{-k_e t}). \tag{25}$$

If the loading dose D_0 is given by injection at the same time that the IV infusion is started at the dosing rate D_r, the concentration function that results is just the sum of these two functions:

$$C(t) = \frac{D_0}{V} e^{-k_e t} + \frac{D_r}{Vk_e}(1 - e^{-k_e t}) \tag{26}$$

or

$$C(t) = C_0 e^{-k_e t} + C_{max}(1 - e^{-k_e t}). \tag{27}$$

Notice that the elimination rate constant k_e is the same for both methods because we are dealing with the same drug—the way the drug is elimi-

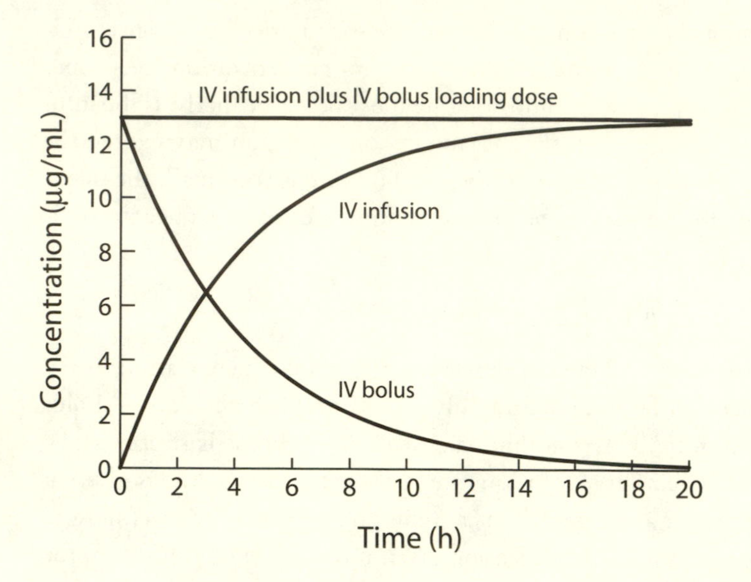

FIGURE 8.19. Concentration function for an IV bolus dose, an IV infusion, and a combination of the two that shows that the concentration is constant.

nated from the body is the same no matter how the drug enters the body and no matter the dosage form.

If the loading dose is chosen so that it equals the dosing rate divided by the elimination rate constant, that is, if $D_0 = D_r/k_e$, then $C_0 = C_{max}$. Thus in equation (27) the two exponential decay terms cancel each other out, resulting in a concentration function that is constant over time, or

$$C(t) = C_{max}. \tag{28}$$

Figure 8.19 shows a plot of the first term of equation (27) for the IV bolus loading dose, a plot of the second term of equation (27) for the IV infusion, and a plot of the combined result showing that the concentration is constant with time.

In the example from the previous section for IV infusion, we had D_r = 15 mg/h and V = 5 L, and k_e = 0.231 h^{-1} ($t_{1/2}$ = 3 h). These values gave us a steady-state (maximum) concentration C_{max} = 13.0 μg/mL. This means that the loading dose must be $D_0 = D_r/k_e$ = 65 mg in order for the concentration to jump to the starting value 13.0 μg/mL and remain constant throughout the time that the infusion takes place.

Oral Dose with Elimination

When a drug is administered orally, the drug enters the bloodstream by passing through the GI wall. As the medication dissolves and is absorbed, the amount of medication in the dose decreases exponentially as a first-order process. In the overall rate equation (12), the absorption rate is written

$$\left(\frac{\Delta C}{\Delta t}\right)_a = k_a C_0 e^{-k_a t}, \tag{29}$$

where C_0 is the total dose of drug D_0 available in the oral medication divided by the volume of distribution V in the bloodstream. The constant k_a is the first-order absorption rate constant. The elimination rate is first order (just as it is for the IV bolus and for IV infusion), so it is written

$$\left(\frac{\Delta C}{\Delta t}\right)_e = -k_e C, \tag{30}$$

where k_e is the elimination rate constant with units of 1/time. Substitution of equations (29) and (30) into equation (12) gives an overall rate equation of the concentration for the oral dosing method of

$$\frac{\Delta C}{\Delta t} = k_a C_0 e^{-k_a t} - k_e C. \tag{31}$$

This differential equation can be solved for $C(t)$ to yield

$$C(t) = C_0 \left(\frac{k_a}{k_a - k_e}\right)(e^{-k_e t} - e^{-k_a t}) \tag{32a}$$

or

$$C(t) = \frac{D_0}{V}\left(\frac{k_a}{k_a - k_e}\right)(e^{-k_e t} - e^{-k_a t}). \tag{32b}$$

This is a fairly complicated, but important, function that combines absorption and elimination rate constants. Perhaps an example using

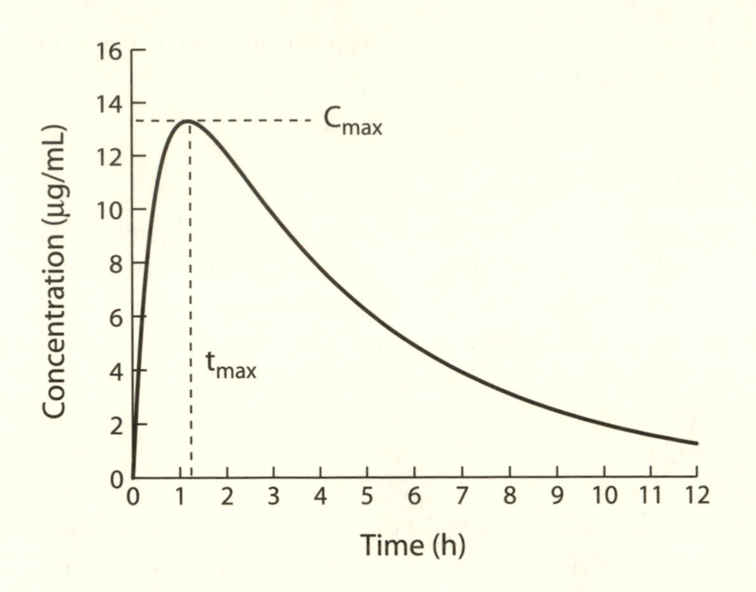

FIGURE 8.20. The concentration function for an oral dose with elimination. It shows a quick rise in concentration initially because the absorption half-life is relatively small. There is a slow decay because the elimination half-life is relatively large. The maximum concentration C_{max} occurs at a point in time called t_{max}.

some actual numbers along with a graph will help us to understand what is happening. Suppose a patient takes a dose D_0 = 350 mg of medication orally, with volume of distribution given by V = 20 L. Assume the absorption and elimination half-lives to be $t_{1/2a}$ = 1/3 h and $t_{1/2e}$ = 3 h, which gives absorption and elimination rate constants of k_a = 2.08 h^{-1} and k_e = 0.23 h^{-1}. In figure 8.20, a plot of the concentration function with time is shown for the parameters given. Notice that the concentration increases rather quickly, reaches a maximum value, and then decreases more slowly. This result is consistent with the smaller value for the absorption half-life than for the elimination half-life.

Usually not all the medication in an oral dose is absorbed and some passes through the GI tract as waste. If this situation occurs, then equations (32a) and (32b) are multiplied by F, the fraction of the dose absorbed. For our discussion, we will assume that all of the medication is absorbed, as is the case with the IV bolus and the IV infusion.

The time t_{max} at which the concentration reaches it maximum value can be shown to be

$$t_{max} = \frac{\ln(k_a / k_e)}{k_a - k_e}.$$

The maximum value of the concentration C_{max} can be found by substituting the value for t_{max} into equations (32) above. For the parameters given, we find $t_{max} = 1.19$ h and $C_{max} = 13.3$ µg/mL, both of which are consistent with the graph in figure 8.20.

If multiple oral doses are given then the concentration is determined in a similar manner as it was for multiple IV bolus doses. The main consideration is that, if there is any drug still in the bloodstream when another dose is given, the starting point for the new concentration function begins at a value determined by the previous oral dose. In figure 8.21, we see the concentration function when a dose is given every four hours. Notice how the maximum values C_{max} and the minimum values C_{min} in-

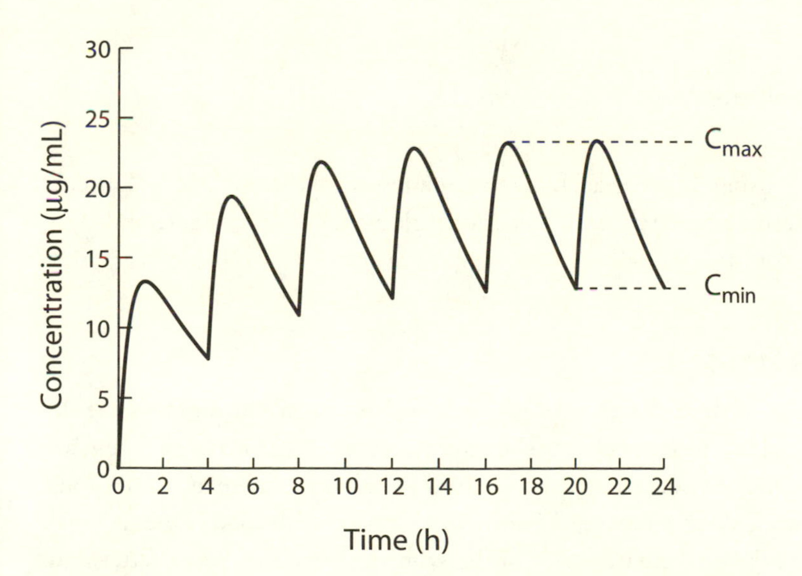

FIGURE 8.21. Multiple oral doses given every four hours show an increase in concentration followed by a decrease during elimination. Steady-state values for C_{max} and C_{min} are reached after about five doses.

crease with time and level out to steady-state values. Once again, these values should lie within the therapeutic window so that the proper clinical response occurs.

The results for multiple oral doses are what one would expect for a person on a regimen of antibiotics for an infection. Suppose a person takes a tablet twice a day, one in the morning and (ideally) another 12 hours later. Over the course of ten days, the concentration varies between the maximum and minimum values and it is hoped that the patient is cured of the illness. Other examples include medication taken for pain or congestion every four to six hours for several days. Perhaps you are all too familiar with the pain or congestion returning before the time interval has passed, tempting you to take another dose before the appointed time. Side effects resulting from too high a concentration may become serious if the frequency of dosing is not followed properly. Also, receptor sites to which the drug is targeted may become saturated so that the higher concentration of drug is not effective. Steady-state maximum concentration levels will depend on the dosing frequency and the amount of drug in the dose.

Applications

Before closing this chapter, let's look at a few applications. We will discuss drug design, time-release medication, blood alcohol content, and drug testing for athletes.

Drug Design

The goal of drug design is to produce a drug form that delivers the desired pharmacological effect. The parameters that go into drug design are many. One consideration is the method of delivery and related questions. Is the drug soluble in water? Should it be taken with food? Does it break down in the acid environment of the stomach? Is liquid form better than solid? What are the properties of the tablet that contains the drug? What about liquid gel formulations? What effect does climate (heat and humidity) have on the shelf-life of the drug?

There is a tremendous amount of research that goes into development of a drug, its packaging, testing, and clinical trials to determine the effectiveness of the drug. Many effects are based on the time-dependent response of the drug in the body, the connection with the pharmacological outcome to the patient, and acceptance of the medication form by the patient. For example, sleeping medication may help a person sleep much better, but what if it has to be given by an IV infusion, it takes several hours to work, and it remains in the body for several days? This is not an appropriate product for someone who needs a pill or tablet that can be easily taken, works quickly, and is almost completely gone from the body in 7 to 8 hours.

Time-Release Medication

One of the most important areas of research is the development of time-release or controlled-release medication. The idea is that if there is an optimal drug concentration for the most effective therapy, delivering that medication in a convenient form will help the patient to achieve the best outcome. You may recall that there is a way to maintain a constant concentration in a patient by using IV infusion with a loading dose given by IV bolus. However, this method is not convenient, nor simple, and can be very expensive. It would be very useful if the medication could be taken orally by the patient when at home or at least away from a clinical setting. Transdermal patches are another method of controlled-release drug delivery.

Research focuses on ways to control the release of medicine in the body. The control can take place in the GI tract, at the target site, or at a barrier that must be crossed. Areas of research include properties of materials, diffusion properties at the target site or at a barrier, and rates of chemical reactions of drugs.

The most common types of rate-controlling materials slowly dissolve over time while releasing medication. Various methods of packaging include encapsulation of the drug in microspheres, placement of the drug in small pockets or micropores, suspension of the drug in a matrix in compressed tablets, or formation of the drug in several layers between layers of inert material. In this last example, the outer inert layer may dissolve quickly so that an initial dose is delivered rapidly, but the inner layer or layers dissolve more slowly, releasing the medication on a delayed time scale. One of the problems with these products is that they may be removed from the GI tract in waste material before they are completely dissolved.

Blood Alcohol Content

Blood alcohol content or concentration (BAC) is a measure of the amount of alcohol in a person's blood. It is a quantity that has become a way of determining the level of intoxication of a person. It is usually measured in grams of alcohol per milliliter of blood, or milliliters of alcohol per milliliter of blood. Because the density of blood is about 1 g/mL, the number is usually expressed as a percentage. Thus, a reading of 0.08% is equivalent to 0.08 grams (or 80 mg) of alcohol in 100 grams of a blood sample. You may be aware that it is illegal in the United States to drive a car if one's BAC is at 0.08% or above.

BAC can be measured in several different ways. The most accurate method is to take a blood sample. Another method measures the amount of alcohol in the breath that comes from evaporation using a breathalyzer; this method provides an estimate of the BAC. Another method measures the alcohol content of urine that results from excretion.

The behavior of a person who has been drinking varies depending on the blood alcohol content. A person with a BAC in the range of 0.01 to 0.06 may feel relaxed or joyful, while being less alert than normal or having impaired judgment. In the range of 0.06 to 0.10%, the person's actions may be uninhibited and reflexes and vision may be significantly impaired. Above 0.11%, there may be wild changes in emotions with much more impaired motor skills. Above 0.30%, the person may experience blackouts and extended periods of unconsciousness, with possible death above 0.40%.

The best treatment of high blood alcohol content is to wait for the alcohol to be removed from the body by excretion, evaporation, or metabolism. Excretion and evaporation account for less than 10% of elimination from the body. The remaining alcohol is metabolized and eventually eliminated. During the time that excess alcohol is being removed, the body responds with a number of undesirable symptoms, including headaches, nausea, and dehydration, to name a few.

Drug Testing

Drug testing is done to determine whether athletes, employees, students, and criminals are using drugs. Athletes use drugs to enhance performance in their respective events. Testosterone and steroids, such as human growth hormone, are injected to build muscle mass to give an athlete an advantage during competition. Alcohol and illegal drugs such as cocaine can affect a person's ability to perform on the job or his or her actions and social behavior. Athletic associations, companies, schools, and law enforcement agencies spend large amounts of money to perform these tests.

Often tests are done looking for specific drugs. Sometimes it is the presence of other chemical substances that indicates a drug has been used because metabolic changes have occurred to the original drug substance. Tests can be done on urine, blood, sweat, saliva, and hair. Each sample requires very careful and specific handling techniques to ensure that the results are accurate.

Technicians use a variety of techniques to determine the presence of drugs. The simplest methods typically involve adding certain liquid chemicals to a sample looking for known chemical reactions to take place. These tests can often be done on site, at a place of business, at school, or at an athletic event. Techniques such as gas chromatography and mass spectroscopy need more sophisticated equipment and can only be done in a laboratory setting.

Often random drug tests are given in order to discourage people from taking unacceptable drugs in the first place. Some drugs can remain in the body for long periods of time. Other drugs may be eliminated much sooner, but their metabolic products may remain in the body much longer.

Summary

The time-dependent behavior of the concentration of drugs in the body is a topic that is widely studied. Physicians, pharmacists, and pharmaceutical scientists need to understand these concepts to properly treat a patient using medication therapy management. Perhaps you have been a bit overwhelmed by all the mathematics in this chapter. There is so much more to this topic that we have not explored, such as the two-compartment model, dose-dependent processes, distribution properties, and metabolism (particularly in the liver). Hopefully, you have begun to appreciate the knowledge and understanding that healthcare professionals need to have as they develop drugs, perform clinical trials, treat specific diseases, and care for patients.

Notes

Chapter One: Motion and Balance

1. Elaine N. Marieb and Katja Hoehn. 2007. *Human Anatomy & Physiology*, 7th ed. 253–276. San Francisco: Pearson Benjamin Cummings.

2. John R. Cameron, James G. Skofronick, and Roderick M. Grant. 1992. *Physics of the Body.* 32–33. Madison: Medical Physics Publishing.

3. Harvard Natural Sciences Lecture Demonstrations. *Vectors and Forces in Equilibrium (Statics)*. http://sciencedemonstrations.fas.harvard.edu.

Chapter Two: Fluids and Pressure

1. Medterms Medical Dictionary. www.medterms.com/script/main/art.asp?articlekey=16163.

2. Ibid. www.medterms.com/script/main/art.asp?articlekey=16164.

3. John R. Cameron, James G. Skofronick, and Roderick M. Grant. 1992. *Physics of the Body.* 95–96. Madison: Medical Physics Publishing.

4. Medline Plus Medical Encyclopedia. *Cerebral Spinal Fluid (CSF) Collection*. www.nlm.nih.gov/medlineplus/ency/article/003428.htm.

5. Cameron, Skofronick, and Grant. *Physics of the Body*, 99–100.

Chapter Three: Energy, Work, and Metabolism

1. T. R. Gowrishankar, Donald A. Stewart, Gregory T. Martin, and James C. Weaver. 2004. Transport lattice models of heat transport in skin with spatially heterogeneous, temperature-dependent perfusion. *BioMedical Engineering On-Line* 3: 42.

2. D. A. Torvi, and J. D. Dale. 1994. A finite element model of skin subjected to a flash fire. *ASME J. Biomech. Eng.* 116: 250–255.

3. K. Giering, I. Lamprecht, O. Minet, and A. Handke. 1995. Determination

of the specific heat capacity of healthy and tumorous human tissue. *Thermochimica Acta* 251: 199–205.

4. Elaine N. Marieb and Katja Hoehn. 2007. *Human Anatomy & Physiology,* 7th ed. 984–985. San Francisco: Pearson Benjamin Cummings.

5. The Physics Fact Book. *Surface Area of Human Skin.* http://hypertextbook.com/facts/2001/IgorFridman.shtml.

6. Steven B. Halls, *Body Surface Area Calculator for Medication Doses.* www.halls.md/body-surface-area/bsa.htm.

Chapter Four: Sound, Speech, and Hearing

1. Jeremy Davey and Dave Mann. *Breaking the Sound Barrier on Land.* www.roadsters.com/750.

2. National Institute on Deafness and Other Communication Disorders. www.nidcd.nih.gov/health/voice.

3. Elaine N. Marieb and Katja Hoehn. 2007. *Human Anatomy & Physiology,* 7th ed. San Francisco: Pearson Benjamin Cummings.

4. Mayo Clinic. April 27, 2007. *Hearing Loss.* www.mayoclinic.com/health/hearing-loss/DS00172.

5. Jerry D. Wilson, Anthony J. Buffa, and Bo Lou. 2010. *College Physics,* 7th ed. 499–500. Upper Saddle River, NJ: Pearson Prentice Hall.

6. Centers for Disease Control and Prevention. *Method for Calculating and Using the Noise Reduction Rating—NRR.* www2a.cdc.gov/hp-devices/pdfs/calculation.pdf.

7. National Institute on Deafness and Other Communication Disorders. www.nidcd.nih.gov/health/hearing.

8. The American Speech-Language-Hearing Association. www.asha.org/public.

Chapter Five: Electrical Properties and Cell Potential

1. Paul Peter Urone. 2001. *College Physics,* 2nd ed. 492–493. Pacific Grove, CA: Brooks/Cole.

2. Elaine N. Marieb and Katja Hoehn. 2007. *Human Anatomy & Physiology,* 7th ed. 388. San Francisco: Pearson Benjamin Cummings.

3. Paul Davidovits,. 2008. *Physics in Biology and Medicine,* 3rd ed. 181–183. Burlington, VT: Academic Press.

4. Marieb and Hoehn, *Human Anatomy & Physiology,* 7: 395–397.

5. Ibid., 75–77.

6. Ibid., 404–408.

7. John R. Cameron, James G. Skofronick, and Roderick M. Grant. 1992. *Physics of the Body.* 184–190. Madison: Medical Physics Publishing.

8. Marieb and Hoehn, *Human Anatomy & Physiology,* 7: 397–409.

9. Morton M. Sternheim and Joseph W. Kane. 1991. *General Physics,* 2nd ed. 455–465. New York: Wiley.

Chapter Six: Optics of the Eye

1. Edward Rosen. 1956. The invention of eyeglasses. *Journal of the History of Medicine and Allied Sciences* 11(1): 13–46; 11(2): 183–218.

2. Nancy N. Schiffer, 2000. *Eyeglass Retrospective: Where Fashion Meets Science.* Altglen: Schiffer.

3. Richard D. Drewry, *What Man Devised that He Might See.* www.teagleoptometry.com/history.htm.

4. Eyetopics. November 24, 2004. *The History of Contact Lenses.* www.eyetopics.com/articles/18/1/The-History-of-Contact-Lenses.html.

Chapter Seven: Biological Effects of Nuclear Radiation

1. Amersham Health. *MYOVIEW kit for the preparation of technetium Tc99m tetrofosmin for injection.* www.amershamhealth-us.com/shared/pdfs/pi/Myoview.pdf.

2. John E. Tansil. October 21, 2000. Natural radioactivity in humans, Presented at the Meeting of the Missouri Association of Physics Teacher, University of Missouri Rolla.

3. National Council on Radiation Protection and Measurement. 1987. *Ionizing Radiation Exposure of the Population of the United States.* Report No. 93.

4. Environmental Protection Agency. June 4, 2008. *Health Effects.* www.epa.gov/rpdweb00/understand/health_effects.html.

5. Mayo Clinic. February 8, 2008. *Radiation Sickness.* www.mayoclinic.com/health/radiation-sickness/DS00432/DSECTION=2.

6. Health Physics Society. February 19, 2008. *Doses from Medical Radiation Sources.* www.hps.org/hpspublications/articles/dosesfrommedicalradiation.html.

7. Bert M. Coursey, and Ravinder Nath. April 2000. Radionuclide therapy. *Physics Today* 53: 25–30.

8. Richard J. Kowalsky and Steven W. Falen. 2004. *Radiopharmaceuticals in Nuclear Pharmacy and Nuclear Medicine,* 2nd ed. 767–788. Washington: American Pharmacists Association.

9. Coursey and Nath, ibid.

10. Amersham Health. *MYOVIEW Kit for the Preparation of Technetium Tc99m Tetrofosmin for Injection*. www.amershamhealth-us.com/shared/pdfs/pi/Myoview.pdf.

Chapter Eight: Drug Delivery and Concentration

1. Leon Shargel, Susanna Wu-Pong, and Andrew B. C. Yu. 2005. *Applied Biopharmaceutics and Pharmacokinetics,* 5th ed. 5–9. New York: McGraw-Hill.

2. Shargel, Wu-Pong, and Yu. ibid. 5–6.

3. Malcolm Rowland and Thomas N. Tozer. 1995. *Clinical Pharmacokinetics: Concepts and Applications*, 3rd ed. 3–5. Baltimore: Williams and Wilkins.

4. For the mathematician, the rate equation is usually written as a differential equation:

$$\frac{dC}{dt}=\left(\frac{dC}{dt}\right)_a+\left(\frac{dC}{dt}\right)_d+\left(\frac{dC}{dt}\right)_e.$$

5. Shargel, Wu-Pong, and Yu. *Applied Biopharmaceutics and Pharmacokinetics,* 5: 9–16.

6. For the mathematician, the rate equation is usually written as a differential equation:

$$\frac{dC}{dt}=\left(\frac{dC}{dt}\right)_a+\left(\frac{dC}{dt}\right)_e.$$

Index

absolute pressure, 38, 39
absolute refractory period, 170
absorption of drug, 246, 249, 254–255, 258, 262–282
absorption rate constant, 263, 270, 279, 280
acceleration, 3, 4; due to gravity, 8
accommodation, 196, 207
acoustic energy, 78, 126, 176
action potential, 166–170
action-reaction forces, 5–6, 139
active processes of charge transport, 151–152, 166
activity, 226–228, 232–234
Adam's apple, 117, 118
air, 182, 183, 197
alcohol in body, evaporation of, 284
alpha decay, 216, 217–218, 223, 224, 232, 235–236, 237, 238
ampere/amp, 153
amplitude, 106, 107, 120, 176
angioplasty, 59–60, 241
angle: importance of, in work, 70–71; of incidence, 183, 184; of refraction, 183
anions, 137
anterior segment of eye, 194
antielectron, 219–220, 235
antimatter, 219, 220
antineutrino, 218–219
aorta, 49, 51, 52, 53, 54, 55, 56, 57
aqueous humor, 62, 63, 193, 195–196, 197, 198
arm, motion of, 15, 16, 26
arrhythmia, cardiac, 172
arteries, 49, 50, 51, 56, 57
astigmatism, 205, 206
atmospheric pressure, 38, 67, 131
atomic mass number, 213–214, 216, 222
atomic number, 213–214, 222
atoms, 136, 137, 140, 210–211, 212, 215, 216, 226
ATP (adenosine triphosphate), 93, 151, 166
audible region, 115–116, 130
auditory canal, 121–122
auditory nerve, 121, 124, 125, 133
auditory tube, 122, 125
auricle/pinna, 121
axis of rotation, 14
axon, 160–163, 167, 169, 170; terminals of, 160–161

Bacon, Roger, 200
balance, 29–30
balloon angioplasty, 59–60, 241
basilar membrane, 123, 124
battery, 146–148, 151, 152, 155, 212
becquerel, 227
Becquerel, Henry, 227
bends, the, 67
Bernoulli effect, 120
beta decay/particles, 216, 218–220, 234–236, 237, 238, 239, 240, 243; beta-minus, 218–219, 223, 224; beta-plus, 218, 219–220, 241
biceps, 6, 16, 25–26
biconcave lens, 186, 188
biconvex lens, 186, 188, 191, 192, 196

bifocals, 204
bioavailability, 246–247
bipolar neurons, 161–162
bladder, 64–65
blood: components of, 195, 251, 252; pressure of, 49–55; speed of, 55–56, 57
blood alcohol content/concentration (BAC), 252, 284
bloodstream, drugs in, 246, 247, 254, 256, 260, 261, 262, 270, 273, 274, 279, 281
body temperature, normal, 88, 97; regulation of, 97–100
boiling point of water, 81, 83
bone mineral density / bone mass density (BMD), 66
bones, 20, 65–66, 118; conduction of sound by, 132
Boyle's law, 39–40, 50, 51, 60, 61
brachytherapy, 240
brain, 63–64, 125, 133, 159, 178; tumors in, 240
brain waves, 172, 252
breathing, 60–61
buccal drugs, 249
bypass surgery, 59–60, 241

californium, 225
Calorie/calorie, 85
calorimetry, 87–88
camera, 181, 184, 189, 198
cancer, 213, 233–234, 239–240
capacitance, 149–152, 167, 170
capacitor, 150–152, 167
capillaries, 49, 56, 57
carbohydrates, 93–94, 101, 102
carbon, 140, 211, 212, 214, 215, 219
cartilage, 21; cricoid, 118–119; thyroid, 118–119
cataracts, 197, 208
cations, 137
cell body, 160–161
cell membrane, 138, 145, 148, 152, 163–170
cell potential, 148, 152, 163–170, 171
center of gravity, 14, 30, 31
central nervous system (CNS), 159–162, 170
cerebral spinal fluid (CSF), 63, 64
channels, gated, 166, 167, 168, 169
charge, electric, 136–139, 143, 144, 145, 146, 149, 175, 176, 214, 235
charge imbalance, 137, 155, 157, 158
charge transport, 151–152, 165–166
charge transport proteins, 151–152
chemical bonds, 140
chemical energy, 77
chemical gradient, 164, 165
chemical reactions, 211
chemical symbol, 213–214
chemotherapy, 246
chlorine ions, 137, 141, 163–164
cholesterol, 59, 60, 94
choroid, 194, 195
cilia, 124, 125
ciliary body, 194, 196
ciliary muscle, 196
ciliary zonules, 196
circuit: pulmonary, 49, 50, 51, 53; systemic, 49, 50, 51, 53, 55
circulatory system, 49, 55
circulatory system, blood flow in, 49–56; arteries, 49, 50, 51, 56, 57; capillaries, 49, 56, 57; veins, 49, 56, 57
circulatory system, drugs in, 246, 247, 254, 256, 260, 261, 262, 270, 273, 274, 279, 281
clinical effect, 246, 251, 254, 259, 261, 275, 282
closed system, 78
cobalt / cobalt treatments, 213, 223, 239
cochlea, 123, 125
cochlear duct, 123, 124
cochlear implant, 133
cochlear nerve, 123
color, 178, 195
color blindness, 208–209
competing process, 221
complex wave, 112, 121
computed tomography (CT), 239, 242
concave lens, 184–188
concentration gradient of ions, 164, 165
condensations, 109, 110, 112, 131
conduction: by bone, of sound, 132; saltatory, 169; thermal, 88, 91–92, 98
conductive keratoplasty (CK), 207

conductivity, 153; electrical, of human tissue, 158
conductor: electric, 157; human body as, 157–158
cones, 195, 208, 209
conservation: of charge, 138–139; of energy, 76, 78–79, 92, 93; of heat energy, 87–88
conservation rules in nuclear decay, 217, 218, 219, 220, 221
contact lenses, 200, 201, 202, 203, 206
controlled-release medication, 283–284
convection, 88, 89–90, 91, 98; forced, 90, 98; natural, 90
convergence, 207
converging lens, 186–188, 189, 190, 203, 204
convex lens, 184–188
cornea, 62, 184, 192, 194, 197, 200, 205–207, 208
coronary arteries, 229, 241, 242
cosmic rays, 238
coulomb, 137
Coulomb force / Coulomb's law, 140–141, 144, 145, 213, 215
curie, 227
Curie, Marie, 227
Curie, Pierre, 227
current, electric, 152–155, 157–158, 162
curvature of lens, 191, 192
cylindrical lens, 205

daughter nucleus, 216–224
decay, exponential: of drug concentration, 253, 258–259, 272, 273, 278; of radioactive nuclei, 226–234, 242–243
decay chain, 222–224
decay constant, 228
decay equation, nuclear, 216–217, 218–226
decay rate, 226–228, 232–234
decibel, 128, 129, 130, 131
decompression sickness, 67
defribrillator, 171, 172–173
dendrites, 160–162
density, 58; linear mass, 108, 119
depolarization, 166–170, 171
deuterium, 215, 226
diabetes, 208, 248
diaphragm muscle, 61, 117
diaphragm, vibrating, 112–113
diastolic pressure, 50
dielectric constant, 150
dielectric permittivity, 150
diffusion, 151, 162, 164, 167, 169
digestive system, 159
diopter, 202
diverging lens, 186–188, 190, 201, 202
Doppler ultrasound, 134
dosage, radiation, 237–239; absorbed, 237; effective, 237–238
dose/dosing rate, 275, 277
dose of drug, 250; loading, 277–278; multiple, 264–265, 269, 273–274, 281–282
drug administration: methods of, 247–250, 258, 260–261, 270; route of, 246, 247; volume of distribution of, 271, 280
drug concentration, 246, 250–255; and compartment models, 262–264, 270; exponential growth of, 253, 258–259, 268; mathematics of, 253–258, 263–282; maximum, 269, 273–274, 275–277, 278, 280, 281; minimum effective, 252, 253, 273, 277; minimum toxic, 252, 253, 274, 277; steady-state, 273, 275, 277, 278, 281–282; using IV bolus, 270–274, 277–278; using IV infusion, 274–279; using oral dose, 279–282
drugs: design of, 282–283; distribution of, 246, 254–255, 258, 261, 262–263; elimination of, 246, 254–255, 258, 261, 262–264, 270–273, 274–277, 284; testing of, 285. *See also* medication

ear canal, 121–22
eardrum, 114, 121, 122, 123, 124; ruptured, 125
earplugs, 129, 130–131, 132
ear trumpets, 132
effector organs, 159–162
efficiency, 79–80, 96
effort, 17
elastic limit, 41

electrical gradient, 165
electric charge, 136–139, 143, 144, 145, 146, 149, 175, 176, 214, 235
electric constant, 140
electric field, 142–144, 149, 157, 175, 176
electric force, 139–141, 143
electricity: in the body, 155–158; static, 155–157, 158
electric potential, 146–148, 153, 154, 156, 163–170
electric potential energy, 144–146, 148, 212, 243
electric properties of neuron, 162–171
electrocardiogram (ECG, EKG), 171, 172, 252
electroencephalogram (EEG), 171, 172
electromagnetic energy, 78, 176, 178
electromagnetic force, 213
electromagnetic spectrum, 176–178
electromagnetic waves, 105, 175–181, 207, 221
electromotive force (EMF), 146
electromyogram (EMG), 171, 172
electron capture, 220–221
electronic configuration, 215, 216, 220–222
electronic transitions, 177
electrons, 136, 140, 153, 154, 157, 176, 177, 178, 210–216, 235
electron-volt (eV), 212, 222
elimination rate constant, 263, 265, 270, 274–275, 277, 278, 279, 280
energy, 69, 72–79, 92, 126, 154; acoustic, 78, 126, 176; chemical, 77; electric potential, 144–146, 148, 212, 243; electromagnetic, 78, 176, 178; from food, 93–94, 101–102; heat, 77, 82–85, 243; internal, 77; kinetic, 72–74, 75, 76, 93, 235; mechanical, 76; nuclear, 78, 211–213, 223, 234–239. *See also* potential energy
energy levels, 215, 216, 220–222
equilibrium, 19, 25, 29–31; mechanical, 19, 25; stable, 29, 30, 31; unstable, 29, 30, 31
equilibrium state of membrane potential, 166
Eustachian tube, 122, 125
evaporation: of alcohol in body, 284; to cool body, 84, 87, 98, 99
excess charge, 137, 155, 157, 158
excretion of drug, 254, 260, 261, 272, 284
exhalation, 60, 61, 117
exponential decay: of drug concentration, 253, 258–259, 272, 273, 278; of radioactive nuclei, 226–234, 242–243
exponential growth of drug concentration, 253, 258–259, 268
eye/eyeball, 61–63, 180, 181, 184, 188, 189, 192, 193–209; aqueous humor of, 62, 63, 193, 195–196, 197, 198; neural layer of, 195; posterior segment of, 194, 196; pupil of, 194, 207; vitreous humor of, 62, 193, 195–196, 197, 198, 208. *See also* optics of the eye; retina
eyeglasses, 181, 184, 188, 200, 201, 202, 204

far point, 199, 200, 201
farsightedness, 202–204, 206
fats, 93, 94, 101, 102
feedback, positive, 169
fever, 100
fibrillation, 157–158, 173
fibrous layer, 193–194
first-order rate constant, 258, 268
first-order reaction/process, 255, 258, 262–264, 265, 267, 268–269, 279
fission, nuclear, 225, 226
flow rate, 44–45, 46, 48, 55, 57, 60
flow rate equation, 44, 55
fluid depth, pressure from, 36
fluid dynamics, 41–48
fluid flow, types of, 42–44
focal length, 189, 191–193, 198–199, 202
focal point, 191
force/forces, 2, 6–12, 70, 176; electric, 12, 139–141; friction, 6, 8–10; net, 3, 4, 72; normal, 7, 10–12; nuclear, 213, 215; tension, 7, 12; weight, 6, 7–8
fovea, 195, 208
fraction: of dose/drug absorbed, 269, 280; of parent isotope remaining, 228, 242–243

Franklin, Benjamin, 200
free-body diagram, 13
free-fall, 8
freezing point of water, 81, 83
frequency, 106, 107, 119–120, 176, 177; fundamental, 108
frequency range of human hearing, 113, 115–116
friction, 6, 8–10, 71; kinetic, 10; static, 10; walking, 9–10
fulcrum, 17
fundamental frequency, 108
fundamental unit of charge, 137, 211, 220
fusion, nuclear, 225–226

gamma decay/rays, 176, 177, 216, 221–222, 223, 226, 234, 235–236, 237–239, 241, 242, 243
Gamma Knife, 240
gastrointestinal (GI) wall/tract, 247, 254, 268, 279, 280, 283, 284
gated channels, 166, 167, 168
geometric optics, 181–184
glands, 159, 233, 240
glaucoma, 61, 62, 196, 208
glial cells, 159, 160
glottal stop, 119
glottis, 119
glucose, 93
graded potential, 167
gravitational potential energy, 74–76, 145, 148
gravity, 8, 213; center of, 14, 30, 31

hair cells, 124, 125
half-life: of drug in bloodstream, 267, 271, 280; of radioactive nuclei, 229–231
Health Physics Society, 239
hearing, 121–133; and auditory nerve, 121, 124, 125, 133; loss of, 124–125
hearing aids, 132–133
heart: blood flow through, 51, 159; chambers of, 51, 171–172; shape of, 53; valves of, 51, 52, 53
heart rate, 173
heat, 77, 82–85, 243; latent, 86–87
heat flow/transfer, 88–92, 98, 100; rate of, 100–101
helium, 211, 217–218, 226
homeostasis, 97
human body: effects of radiation on, 213, 234–239, 243; specific heat of, 86, 100. *See also specific body parts*
hydrocephalus, 64
hydrogen, 211, 212, 215, 225, 226
hyperopia/hypermetropia, 202–204, 206
hyperpolarization, 166–170
hyperthermia, 100
hyperthyroidism, 233
hypothermia, 100

ideal gas law, 39, 109
image: real, 188–190, 196, 201; resolution of, 134; virtual, 188–190, 201, 203
image distance, 193, 198
image formation, 194, 198, 199
incontinence, 61
index of refraction, 182–184, 197–198
inertia, 2
infrared (IR) radiation, 176
infrasonic region, 115–116
infusion: intravenous, 57–58, 249, 250, 254, 261, 262, 264–266, 270, 274–278; rate of, 249, 265, 270, 274–275, 277
inhalation: of air, 60, 61, 117; of drug, 250
injection: intramuscular, 248, 249; intravenous, 248, 250, 254, 261, 262, 264, 265, 270–274, 277–278; subcutaneous, 248–249
inner ear, 121, 123–124, 125
insertion point of muscle, 21
insulators, electrical, 137, 153, 155, 160, 167
insulin, 248
integration, 159
intensity, sound, 126–131
intensity level, sound, 128–131
internal energy, 77
interneurons, 161–162
intracranial pressure, 63–64
intramuscular injection, 248, 249
intraocular pressure, 61–62, 196, 208

intravenous (IV) bolus or injection, 248, 250, 254, 261, 262, 264, 265, 270–274, 277–278
intravenous (IV) infusion: of drug, 249, 250, 254, 261, 262, 264–266, 270, 274–278; and pressure, 57–58
inverse-square law, 128
iodine, 233
ionic solution, 138
ionizing radiation, 210, 234, 237
ions, 137, 141, 153
iris, 194, 207
irradiation of food, 243
isotopes, 211, 214–216, 222. *See also* radioisotopes

joints, 21, 32, 65; synovial, 21

Kelvin, 80, 81, 82
kinetic energy, 72–74, 75, 76, 93, 235

laminar flow, 43, 55
larynx / voice box, 117–118, 121
LASEK (laser epithelial keratomileusis), 206
LASIK (laser-assisted intrastromal keratomileusis), 205–206
latent heat, 86–87
law: of charges, 213; of motion, 2–6, 70, 72, 77, 139, 141, 143; of refraction, 183
lead, 215, 223, 235
Leonardo da Vinci, 200
lens, 180, 184–193; biconcave, 186, 188; biconvex, 186, 188, 191, 192, 196; concave, 184–188; contact, 200, 201, 202, 203, 206; converging, 186–188, 189, 190, 203, 204; convex, 184–188; cylindrical, 205; diverging, 186–188, 190, 201, 202; of the eye, 62, 194, 196–199, 207, 208; spherical, 184–193
lens implants, 207
lens maker's equation, 191
lever arm, 15
levers, 17, 23–24
ligaments, 118, 194, 196, 204
light, 179–180; speed of, 176, 177, 181–183, 235; visible, 176, 177–178, 195
linear mass density, 108, 119
lipids, 93, 94, 101, 102
lithotripsy, 134
load, and levers, 17
loading dose, 277–278
longitudinal wave, 105–106, 110
loudspeaker, 112–113, 128
lungs, 60–61, 117, 118, 120
lymphoma, non-Hodgkin's, 240

macula, 195, 208
macular degeneration, 208
magnetic field, 175, 176
magnifying glass, 184, 189, 191, 200, 204
mass, 3, 136–137, 176; of radiation, 235; of radioactive sample, 231–232, 233
mass-spring system, 108
mechanical advantage, 18
mechanical energy, 76
mechanical equilibrium, 19, 25, 26, 29, 30, 31
medication: oral, 247–248, 250, 254, 261, 262, 268–269, 279–282; rectal, 249; sublingual, 249; time-release, 283–284
membrane, vibrating, 110, 112, 114
membrane potential, 148, 152, 163–170, 171
meninges, 63
metabolic rate, 96, 101, 102; basal, 96; total, 96–97
metabolism, 96–97; of drugs, 254, 260, 261, 271–272, 284
metals, 137
metastability, 222
microkeratome, 206
microphone, 114, 132, 133
microscope, 181, 184, 189, 193
microwaves, 176
micturition pressure, 65
middle ear, 121, 122–123, 124, 125
minimum effective concentration (MEC), 252, 253, 273, 277
minimum toxic concentration (MTC), 252, 253, 274, 277
motion, 2, 13, 14, 22, 23, 31, 32; laws of, 2–6, 70, 72, 77, 139, 141, 143; ro-

tational, 14; translational, 13; walking, 31–32. *See also specific body parts*
motor neurons/nerves, 159–162
motor output, 159–162
mouth, 121
movement, difficulty of, 32
movement of body part: angular, 23; gliding, 23; rotational, 23
mucous membranes, 249–250
multipolar neurons, 161–162
muscle fibers, 20
muscle tension, 22; contraction, 22, 157, 159, 161, 172; isotonic, 22
muscles: biceps, 6, 16, 25–26; cardiac, 19–20; eye, 191, 194, 207; deltoid, 24, 26–27; erector spinae, 28; gastrocnemius, 24; larynx, 118; neck, 7; origin and insertion point of, 21; skeletal, 19–20, 97, 98; smooth, 20
musical instruments, 107, 108, 116
myelinated axons, 169
myelin sheaths, 160, 162
myopia, 200–202, 204, 205, 206
MYOVIEW, 228, 242–243

nasal cavity, 121
nasal drugs, 250
near point, 199
nearsightedness, 200–202, 204, 205, 206
nerve cell. *See* neuron
nerve endings, 160–161
nerve impulse, 166–170
nervous system, 158–171
neuroglia, 159, 160
neuron, 138, 148, 149, 159–171, 172; bipolar, 161–162; electric properties of, 162–171; motor, 159–162; multipolar, 161–162; sensory, 159–162; signal transmission in, 160, 162, 166–170; unipolar, 161–162
neurotransmitters, 160
neutrino, 218–221, 225, 226
neutron activation, 225
neutrons, 136, 210–216, 225, 226, 235; emission of, 225
Newton's laws of motion, 2–6; first law, 2, 70, 72; second law, 3, 141; third law, 5, 77, 139, 143
nitrogen, 219, 220, 225
nodes of Ranvier, 162, 169
noise, 131–132
noise-canceling headphones, 129, 130–131, 132
noise reduction rating (NRR), 130
normal force, 7, 10–12
normal, 182
nuclear decay equation, 216–217, 218–226
nuclear energy, 78, 211–213, 223, 234–239
nuclear force, strong, 213, 215
nuclear notation, 213–214, 217–226
nuclear radiation, 210, 215, 216–217, 222–225; and decay rate, 227–229; effects of, in humans, 213, 234–239; and half-life, 229–231; medical applications of, 239–243. *See also* alpha decay; beta decay; gamma decay
nucleons, 211, 213, 215, 216, 217, 219
nucleus, 137, 140, 141, 210–214, 216–224, 226; daughter, 216–224; parent, 216–224, 227, 228, 229, 231; stability of, 214–216, 223, 224; unstable, 214–224

object distance, 193, 198
ohm, 153
Ohm's law, 153–154, 157, 158
ointments, topical, 250
one-compartment model, 259–261, 262, 263–264
open system, 78
optical devices, 181, 184, 189, 198
optic disk, 195
optic nerve, 62, 63, 178, 195, 208
optics of the eye, 175, 178, 181, 184, 188, 192, 207–209; geometric optics, 181–184; and path of light, 193–198. *See also* astigmatism; hyperopia; LASIK; myopia; presbyopia; vision: normal
oral dose/medication, 247–248, 250, 254, 261, 262, 268–269, 279–282
origin of muscle, 21
ossicles, 122, 123, 124, 125; incus/anvil, 123; malleus/hammer, 123; stapes/stirrup, 123

osteopenia, 66
osteoporosis, 66
otitis media, 122, 125
otosclerosis, 125
outer ear, 121–122, 124
oval window, 123

palliative agent, 240
parallel processing, 171
parent nucleus, 216–224, 227, 228, 229, 231
Pascal's principle, 37–38, 50, 51, 58, 65
passive processes of charge transport, 151–152, 165
period, 106, 107
periodic table, 219, 220
periodic waves, 104, 105, 106
peripheral nervous system (PNS), 159–162, 170
peripheral vision, 195
permeability, 164
permittivity of free space, 150
perpendicular, 182
perspiration, 84
PET scan, 241
pharmacokinetics, 246, 252, 261, 269, 274
pharmacological effect, 246, 251, 254, 259, 261, 275, 282
pharynx, 121, 125
photons, 216, 222, 223
photoreceptors, 195
photorefractive keratectomy (PRK), 205–206
physiology of speech and hearing, 116–124
piano, 108, 116
pitch, 107, 119–120
pivot point, 14
Planck's constant, 176
planetary model of atom, 211
plasma, blood, 195, 251, 252
plasma level–time graph, 252–253, 264
platelets, 251
plutonium, 217–218
point source: of light, 179–180; of sound, 127–128, 179
Poiseuille's law, 46–48, 49, 57, 59
polarization, 166–170
polonium, 232, 236
positron, 219–220, 235
potassium ions, 137, 141, 163–166, 168
potential difference, 147, 171
potential energy, 72, 74–76, 93, 94, 144; elastic, 74; electric, 75, 77, 144–146, 148; gravitational, 74–76, 145, 148
power, 79, 80, 126; electric, 154, 155; of lens, 189, 192, 202, 203
pregnancy, 65, 101
presbyopia, 204, 206
pressure, 34–41, 120, 124; absolute, 38, 39; air, 38, 120, 124, 131; atmospheric, 38, 67, 131; blood, 49–55; diastolic and systolic, 50; in eye, 61–63; intracranial, 63–64; intraocular, 61–62, 196, 208; and IV infusion, 57–58; in lungs, 60–61; micturition, 65; in solids, 40–41; sound, 131; units of, 35
pressure, causes of: external force, 37–38, 46–48, 49, 57; interaction of molecules, 38–40; weight of fluid, 35–37, 53–54, 58, 67
pressure gauge, 38, 50
pressure gradient, 48
pressure wave, 109–111, 123, 131
prism, 184, 185
protein, dietary, 93, 94
proteins, charge transport, 151–152
protons, 136, 140, 153, 210–216, 219, 235; emission of, 225
pulmonary artery, 51
pulmonary circuit, 49, 50, 51, 53
pupil, 194, 207
Purkinje fibers, 172
P wave, 172

quantization of charge, 138
quantum physics, 211
quinacrine, 252

rad (radiation absorbed dose), 237–239
radial keratotomy (RK), 205
radiation: infrared and ultraviolet, 88, 90, 91, 98. *See also* nuclear radiation
radiation sickness, 238
radioactive dating, 215

radioactive decay, 215, 224, 226–234
radioimmunotherapy (RIT), 240
radioisotopes/radionuclides, 215, 216, 219, 226, 228, 230, 232, 235, 240, 241
radio waves, 176, 177
radium, 218
radon, 218, 232, 238
rarefactions, 109–110, 112, 131
rays, 178, 179, 180–181
real image(s), 188–190, 196, 201
red blood cells, 251
reference point, 147
reflection, 180–184
reflexes, 171
refraction, 180–185, 192, 194, 200, 202, 205, 207; index of, 182–184, 197–198; law of, 183
relative biological effectiveness (RBE), 237–238
relative refractory period, 170
rem (radiation effective man), 237–239
repolarization, 167–170, 171
resistance, 152–155, 170
resistivity, 153
resistor, 153
resolution of image, 134
resting potential, 166, 167
retina, 62, 188, 189, 192, 195, 196, 197, 198, 200, 202, 203, 208
rods, 195
rotational motion, 14

sacrum, 28
saltatory conduction, 169
Schwann cells, 162
sclera, 194
seed therapy, 240
semicircular canals, 123
semiconductor, 153
senses, 161–162. *See also* hearing; vision
sensory input, 159–162
sensory layer, 193–195
sensory neurons/nerves, 159–162
sensory receptors, 159–162
serial processing, 171
shell model of atom, 211
shielding, radiation, 236, 242
shock, electric, 156, 158, 159
signal transmission in neurons, 160, 162, 166–170
simple harmonic motion (SHM), 106–109, 176
sinoatrial (SA) node, 172–173
skull, 63, 64
Snell's law, 183
sodium iodide, 232–233
sodium ions, 141, 142, 163–166, 168–169
sodium/potassium pump, 149, 152, 166, 169
sound, 103–135; detection of, 114–116; point source of, 127–128, 179; production of, 112–113, 116–121
sound barrier, breaking of, 111–112
sound frequency spectrum, 115–116
sound pressure level (SPL), 131
sound waves, 109, 110, 177; complex, 112, 121; simple, 112
specific heat, 85–87; of body, 86, 100
SPECT, 241
spectrum: electromagnetic, 176–178; sound frequency, 115–116; visible light, 176, 177–178, 195
speech and hearing, 116–124
speed: of blood, 55–56, 57; of light, 176, 177, 181–183, 235; of nerve impulse, 169
speed of sound, 107, 111–112, 116; in air, 111–112, 116; in human tissue, 134
spherical aberration, 194, 207
spherical lens, 184–193
sphygmomanometer, 50, 54
spinal cord, 159
spine, 28
spiral organ of Corti, 124
stable equilibrium, 29, 30, 31
standing waves, 104–105
static electricity, 155–157, 158
stenosis, 241
stent, 241
stimulus, 166–170
strain, 41
stress, 40, 65
subcutaneous injection, 248–249
sublingual drugs, 249
swimmer's ear, 125

synapses, 160, 172
systemic circuit, 49, 50, 51, 53, 55
systolic pressure, 50

tachycardia, 173
technetium, 222, 229, 242–243
telescope, 189, 193
temperature, 80–81; body, 88, 97–100; change in, 81–82
tendons, 20–21
tension, 12; in ligaments, 196, 204; in string, 108; in vocal cords, 119
test charge, 142
Tetrofosmin, 242–243
therapeutic nuclear radiation, 239–241
therapeutic range/window, 252, 253, 273
thermal conductivity, 91, 92; of body, 100
thermal contact, 82, 87
thermal equilibrium, 82, 87
thermal processes of atoms, 211
thin lens equation, 192–193, 198
thoracic cavity, 61, 117
threshold: of hearing, 126, 128, 129, 131; of pain, 126, 129, 131
threshold potential, 168
thyroid gland / thyroid cancer, 233, 240
time-release medication, 283–284
timolol, 208
tomography, 241–242
tonometer, 62
torque, 13, 17–19
trachea, 61, 117, 118
transdermal patches, 250, 283
translational motion, 13
transparent media/materials, 181, 183
transverse wave, 105–106
trifocals, 204
triglycerides, 94
tritium, 215
turbulence, 43, 54
TV waves, 176, 177
T wave, 172
tweeters, 113
two-compartment model, 260, 261–262, 263
tympanic membrane, 114, 121, 122, 123, 124; ruptured, 125

ultrasonic region, 115
ultrasound imaging, 115, 133–134
ultraviolet (UV) radiation/rays, 176, 177, 210
unipolar neurons, 161–162
units: of activity, 227; of current, 153; of electric charge, 137, 211, 220; of electric potential, 146; of energy/heat, 74, 84–85; of radiation dosage, 237–239; of resistance, 153; of sound intensity and intensity level, 126, 128, 131
unstable equilibrium, 29, 30, 31
uranium, 212, 215, 218, 223, 232
urinary retention, 43
urination, 65
uvea, 194

vacuum, 182
Van de Graaff generator, 156, 157
vascular layer, 193–194
vector, 13
veins, 49, 56, 57
vena cava, 49, 56, 57
ventricles, 51, 53
vestibular folds, 120
vestibule, 123
vibrations, 106, 124
virtual image, 188–190, 201, 203
viscosity, 42, 45–48, 57, 60
visible light/spectrum, 176, 177–178, 195
vision: normal, 195, 199, 200, 204; problems with, 199–209
vitamins, 93, 94
vitreous humor, 62, 193, 195–196, 197, 198, 208
vocal cords/folds, 108–114, 116–121
voice, 116, 120
volt, 146
voltage, 146, 153

walking, 9–10, 31–32
water: index of refraction of; 182, 183, 198; latent heat of fusion of, 86; latent heat of vaporization of, 86–87; specific heat of, 85
wave front, 179–181
wavelength, 107, 176, 177, 178, 179, 195

waves, 103–106, 110, 178–181; brain, 172, 252; complex, 112, 121; electromagnetic, 105, 175–181, 207, 221; longitudinal, 105–106, 110; microwaves, 176; periodic, 104, 105, 106; pressure, 109–111, 123, 131; radio and TV, 176, 177; sound, 109, 110, 112, 121, 177; standing, 104–105; T, 172; transverse, 105–106
weight, 6–9, 14–15; body, gain/loss of, 101–102
white blood cells, 251
white noise, 131–132
woofers, 113
work, 69–72, 74, 75, 78, 92, 94–96, 144, 149; net, 71, 72, 73
work-energy theorem, 74

x-rays, 176, 177, 210, 221, 234, 235–237, 238, 239, 241, 242, 243

zero-order rate constant, 256
zero-order reaction/process, 255–257, 262, 264–265, 267